Floating Point Numerics for Games and Simulations

Floating point is ubiquitous in computers, where it is the default way to represent non-integer numbers. However, few people understand it. We all see weird behavior sometimes, and many programmers treat it as a mystical and imprecise system of math that just works until it sometimes doesn't. We hear that we shouldn't trust floating point with money, we know that 0.1 + 0.2 does not equal 0.3, and "NaN" shows up in our logs when things break. We rarely hear why any of this is the case, and less about what to do about it.

This book pulls back the veil on floating point and shows how this number system we program with every day works. It discusses how to leverage the number system for common calculations, particularly in graphics and simulations, and avoid pitfalls. Further, we will review methods that can give you either better performance or better accuracy on tasks like numerical integration and function approximation, so you can learn to make the right tradeoffs in your programs.

This book builds upon a basic knowledge of calculus and linear algebra, working with illustrative examples that demonstrate concepts rather than relying on theoretical proofs. Along the way, we will learn why Minecraft has struggled with boat physics and what the heck John Carmack was thinking with *Quake III*'s infamous fast reciprocal square root algorithm. By the end of the book, you will be able to understand how to work with floating point in a practical sense, from tracking down and preventing error in small calculations to choosing numerical building blocks for complex 3D simulations.

- Gives insight into how and why floating-point math works
- Describes how floating-point error arises and how to avoid it
- Surveys numerical methods important to graphics and numerical simulations
- Includes modern techniques to apply to your numerical problems
- Shows how to hack the floating-point numbers to compute faster and more accurately

Nima Badizadegan is an engineer who works at the intersection of computer systems and mathematics. His past experience includes work at Google and on Wall Street, as well as being a consultant and startup founder. Badizadegan has several publications in the fields of simulation and computer arithmetic and is the inventor of over 10 patents. He is a member of the IEEE and ACM and contributes to the IEEE 754 floating point standard. He is the author of the popular technical blog Speculative Branches (https://specbranch.com), where he writes about computer systems, software engineering, and math.

Floating Point Numerics for Games and Simulations

Nima Badizadegan

CRC Press
Taylor & Francis Group
Boca Raton London New York

CRC Press is an imprint of the
Taylor & Francis Group, an **informa** business

Designed cover image: Nima Badizadegan

First edition published 2025
by CRC Press
2385 Executive Center Drive, Suite 320, Boca Raton, FL 33431, U.S.A.

and by CRC Press
4 Park Square, Milton Park, Abingdon, Oxon, OX14 4RN

CRC Press is an imprint of Taylor & Francis Group, LLC

ISBN: 978-1-032-93356-6 (hbk)
ISBN: 978-1-032-93355-9 (pbk)
ISBN: 978-1-003-56554-3 (ebk)

DOI: 10.1201/9781003565543

Typeset in CMR10 font
by KnowledgeWorks Global Ltd.

Publisher's note: This book has been prepared from camera-ready copy provided by the authors.

This book is dedicated to three math teachers: Mr. Brian Marks, whose math classes inspired me as a child, Professor Isaac Fried, whose numerical analysis course started my exploration of the field, and Mr. Nick Dent, who periodically shares difficult math problems with me—the problems in this book are for you.

Contents

Preface

There are two default ways to represent numbers on computers: Integer and floating point. If you work with numbers on computers, floating point quietly underlies everything you do. Games and simulators fit at the center of this world, building experiences and experiments that are backed by floating point. Most of the time, it works flawlessly, but every once in a while, the abstraction leaks.

The best-selling video game in history is Minecraft. In Minecraft, when you fall off a high cliff while your character is in a boat, you will not take fall damage. That is, unless you fall exactly 12, 13, 49, 51, 111, 114, 198, 202, 310, or 315 blocks. From those heights, there has been a persistent bug that causes your boat to break. The player character will die in a way that defies the "rules" of the game, with seemingly no explanation. You can thank floating-point numerics for that death. This bug, reported in 2017 and still persistent as of 2024 in some cases despite workarounds, comes from the interaction of two pieces of numerical code deep in Minecraft's game engine.

Floating point is so successful as a number system that several programming languages use floating-point numbers as their default numeric type. The most popular programming language today, Javascript, follows this pattern. It is not an exaggeration to say that every programmer in the world has interacted with and relied on floating point.

I first came to seriously thinking about and contributing to the IEEE 754 floating point standard in 2017, right at the end of the process of designing and ratifying the most recent version of the standard (yes, the *floating point* standard's most recent update came in 2019). At the time of writing, this is the most recent version of the standard. Although many of the core features of floating point like addition and multiplication have been stable for several decades, some features, like fused-multiply add operations, are relatively new. Beneath you work, computer arithmetic is alive and advancing.

As part of the process of working in computer arithmetic and the IEEE 754 standard, I realized that there was a need for a user's guide, that takes the abstract ideas involved in creating the standard and makes them more concrete and usable. As such, this book is intended to be a "field guide" to the floating-point number system and numerical analysis for practitioners like me. It is intended to be approachable to you without mathematical background beyond basic calculus and linear algebra. Instead of the theorems and proofs you might find in a numerical analysis book, we will be taking a look at illustrative examples and case studies, since much of the theory of numerical analysis relies on continuous arithmetic, anyway. Thus, this book is an attempt to cover the practice of numerics on computers.

The first two chapters of this book introduce the floating point numbers as an approximation of the real numbers. Chapters 3–8 discuss floating point operators, beginning with an introduction of the universal features of every operation in Chapter 3, rounding and exceptions. Chapter 4 discusses the basic arithmetic operators, Chapters 5 and 6 cover the operations that are most likely to realize problems from error: Comparisons and conversions. Chapter 7 covers the transcendental functions you are likely to see, and Chapter 8 covers several useful operators which help you bend the floating-point abstraction.

Chapters 9–12 discuss numerical analysis in computing. We begin with a discussion of methods of handling imprecision in calculation chains in Chapter 9, with Chapter 10 discussing function approximation and the methods that underlie the functions from Chapter 7, so you can build your own functions. Chapter 11 discusses numerics in geometry in the context of a discrete number system. Finally, we conclude with a discussion of numerical integration in floating point in Chapter 12, one of the foundational pieces of simulators.

While the latter half of this book depends on what some might call heavy mathematics, I have kept the number of formal proofs to a minimum, instead preferring to work with motivating examples. For this reason, the bibliography skews toward books and papers that are deeply technical and mathematical for those who are interested in a more formal and rigorous approach to any specific subject. I have also marked the formal derivations in this book with the use of Greek letters, so you can read closely or avoid those sections altogether based on your taste.

In the cases of descriptions of hardware, compilers, and historical events, the bias is toward primary sources, while other resources may have better explanations. Additionally, there is some bias toward IEEE and ACM papers as well as books published in the United States, because that is what you will find in the author's personal library. The IEEE 754 standard [1] itself is used quite extensively as a reference, so while it is reference number one in the bibliography, you will not find it cited in the text, as there would be too many citations to be useful.

Finally, every chapter includes a set of exercises. Several of these exercises are intended to be deep (and sometimes tricky) applications and expansions of the material from the chapter, and they are also intended to be ordered by difficulty. Some are open-ended and some are motivated by recent talks and research papers. One problem from Chapter 2 inspired a paper of its own. It is not out of the question for one of the later problems in any chapter to be a 10-hour adventure, but I hope you find the time rewarding.

Figures

Tables

Algorithms

1

Computing with Numbers

Ever since the invention of computers, humans have used computers to do arithmetic. The first computer, ENIAC, was built to compute artillery tables, and ended up simulating nuclear explosions. The history of numerical computing is the history of computing. This also means that the legacy of numerical computing, like it or not, is with us today.

I like to joke that computers were built for video games, although this is not exactly true. They, and the number formats they use, were built for war-related physics simulations. Artillery tables and calculation of jump trajectories use the same underlying math. Collisions, explosions, and other common features of games are also strategically important to understand when you reframe them as collisions of atomic nuclei or yield calculations for bombs. The math that built weapons of war now gives us the power to understand the world, and has produced the most popular medium of entertainment of the 21st century.

From the beginning, computers have had to contend with the fact that space is continuous, while bits are discrete. Any representation of that continuous space is guaranteed to be a compromise of some kind, trading off computational complexity for the ability to represent pieces of the range of real numbers. Computers also rely on software, and a further challenge is the ability to compose software around your number system, bringing about a third dimension on which to compromise: The complexity to the developer.

Among the number of compromises that you can accept when computing with numbers, the compromises of floating point have the power of giving good results while also avoiding mental load on the programmer and being reasonably efficient to compute. Floating-point arithmetic actually predates the computer. The first machines using floating point were built in 1938, preceding ENIAC, the first computer, by several years [2]. It was found to be a format that balanced the many compromises involved in representing the real numbers in a computable way. The format has continued to be a mainstay for numerical computations due to its wide dynamic range and stable precision, eventually becoming the default option.

DOI: 10.1201/9781003565543-1

1.1   Between the Integers

Digital computers are discrete machines. There is no question of how to represent a positive integer on a digital computer, and there is only one good way to do it: With binary. The negative integers are a bit more interesting, but the two's complement format became the de-facto standard way to implement negative integers due to the simplicity and ease of operations. If we want to represent numbers that sit between the integers, however, there are still many possible number formats, and all have their pluses and minuses.

The integers are a small set of numbers compared to the full set of real numbers we use to calculate. Between each of the integers, there are an infinite number of numbers. Splitting the difference between 0 and 1, we get $\frac{1}{2}$. We can split the difference again to get $\frac{1}{4}$, and so on. These lead us to the rational numbers. Rational numbers are also representable in computer systems as a pair of integers, and have some niche applications. Arithmetic on rational numbers usually requires reducing the fraction to avoid magnitude problems:

$$\frac{1\,268\,609\,872\,658\,765}{2\,537\,219\,745\,317\,530} = \frac{1}{2}$$

Generic arithmetic operations on fractions produce a denominator that is the product of the denominators of the two operands, and this means that algorithms for operating on rational numbers must reduce the fraction frequently. Reducing fractions is a computationally intensive problem, since it requires finding the greatest common divisor of the numerator and denominator, two numbers which we expect to be large any time we are doing a reduction. This makes rational numbers unattractive for numerical computing.

Stepping back to the land of mathematics, the rational numbers still have gaps. They do not account for numbers with an infinitely long non-repeating decimal representation. Irrational numbers like e and π have no fraction that can represent them. These gaps are only filled by the real numbers, and it is the real numbers where most of mathematics actually takes place. We have no hope of actually computing efficiently on the real numbers, but our task is doing better than the integers and rational numbers.

1.2   Representing the Real Numbers

The real numbers are rigorously defined as the one-dimensional field that is both continuous and complete. When we look at a number line, like Figure 1.1, we usually put ticks on the integers, but the real numbers are the line. Continuous means that any pair of real numbers can be arbitrarily close together and that any point along that line is a valid real number. Completeness means

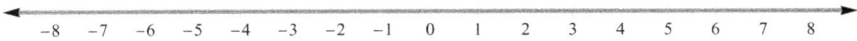

FIGURE 1.1
A number line of the real numbers shown in the range $[-8, 8]$ with ticks at the integers.

that there are no possible gaps in the set: The number line of the real numbers can be drawn with one infinite stroke of an infinitesimally small pen. These properties make the real numbers ideal for mathematics and physics, but terrible for computing.

Our task when constructing computer number systems for real arithmetic is to try to get close to the real numbers. Every real number has a decimal representation (some, like $0.9999\ldots = 1$, have two), but those representations can be infinitely long. On a computer, we have finite memory, so we can't store real numbers in their possibly infinite representations. We would prefer that our numbers have a fixed bit width, too. This inevitably means quantizing the space in which we represent the numbers. We have to find points on the number line to represent and accept that we cannot represent the rest of them exactly. The number of points we can represent is determined by the size of numbers we store. Quantization means that we cannot have a number system that is either continuous or complete, and we have to fake those properties if we want our numbers to look like the reals. In that sense, floating-point numbers are mathematically more like integers than like real numbers.

Additionally, when we quantize our numbers, we lose the scope of the real numbers. The numbers $10^{-100000000}$ and $2^{999999999999999}$ are both in the real numbers. When we quantize our number space and pick a finite number of them to represent, we must deal with the fact that any quantization of the real numbers probably cannot include both of those two examples (and any sane quantization will contain neither of them).

An alternative possibility is that we decide to keep numerical accuracy instead of having a fixed size. This decision results in one of two solutions: Arbitrary-precision arithmetic, where numbers are stored using as much memory as required, up to the full precision of the machine, or symbolic computation, where the computer directly does real-number algebra and only evaluates (with arbitrary precision) at the end of the calculation. The former is used by calculators and calculator applications, while the latter is the domain of computer algebra systems, like Mathematica. These number formats are incredibly computationally intensive, and operations often take non-constant time. For any situation where performance is of interest, these number systems are difficult to justify.

There was another word in the definition of the real numbers that is load-bearing: "Field". A field isn't just a set of numbers. It is a set of numbers that is paired with two operators addition and multiplication, so that you can do algebra with those operators and so that every operation gives you a number in the field. Finally, fields contain identity element for each operator, 0 for addition and 1 for multiplication, which is the element of the set that gives you back the number you added or multiplied with it, and contains all inverses of all elements—the negative numbers are the additive inverses of the positive numbers and $\frac{1}{x}$ is the multiplicative inverse of x. This definition has a lot of requirements that are also hard to satisfy with quantization. For example, it is hard to have inexact mathematical operations satisfy the algebraic properties required. Even the integers cannot form a mathematical field, since the multiplicative inverses of the integers are not integers.

We are left with having to make a series of compromises when we want to construct an arithmetic format that is practical for computing. At the core of those compromises is quantization: The move from a continuous, infinite space to a discrete, finite set. Quantization means that when we do math, we need to consider the properties of the number space itself in addition to the properties of the math we want to do. In a sense, the "miracle" of floating point is that we rarely ever feel the effects of quantization on calculations.

1.3   Properties of Number Formats

Number formats on computers are compromises, and they have several important properties that allow us to compare them. The two major properties we will consider are precision and dynamic range, which in turn tell us how fine-grained our quantization is and how much total length on the number line we can represent.

Precision indicates how large of a range of the real numbers maps to each number in our quantized number system, and is often specified relative to the magnitude of a number to put it in terms of a common concept in science: Significant digits. Precision is a measure of how many significant figures we can possibly represent. For example, the integer number 15 maps to all of the real numbers between 14.5 and 15.5. The maximum absolute error between the real value and the quantized value is 0.5, so this number is about 3.3% precise, which corresponds to about four bits of precision. A mathematical definition of precision is:

$$P = \max_{r \to q} \left| \frac{r - q}{q} \right|$$

where r is a real number that maps to the given quantized number q. When we take $- \log_2$ of precision, we get the maximum number of bits of significance that a number can have, and when we take $- \log_{10}$ we get the number of significant figures. The calculation for bits of precision is:

$$P_b = - \log_2 \left(\max_{r \to q} \left| \frac{r - q}{q} \right| \right)$$

Significance carries through a computation. When we multiply the integer 15 by 1 000, the resulting number has gained precision, since 15 000.5 and 14 999.5 are closer in relative terms to 15 000 than 14.5 is to 15, but the result of the calculation cannot gain significance. Since 14.5 and 15.5 could have mapped to 15, the range of real numbers that corresponds to the result of $15 \times 1\,000$ (assuming the 1 000 is exact) is 14 500 to 15 500. The error we acquire from representing an intermediate calculation result with the number "15" carries through our full chain of operations.

You may have guessed by now that when the integers are used for real-number calculations, they have varying precision with their magnitude. "100 000 000" is fantastically precise while "4" is imprecise. This makes numerical analysis in the integers quite an adventure, as any operation that causes a decrease in magnitude causes a corresponding significance loss that is stuck with you forever.

The second property is dynamic range. The dynamic range of a number system is the ratio between the largest-magnitude and smallest-magnitude numbers that can be represented in our quantized set. Dynamic range is a term adapted from audio and signal processing, where it characterizes the difference between the loudest possible sound you can hear and the quietest possible sound you can hear. In equation form, the dynamic range of our quantized set Q is:

$$DR = \max_{q_1 \in Q} |q_1| \, / \, \min_{q_2 \in Q} |q_2|$$

Having a wide dynamic range gives a number system the ability to represent very large and very small numbers simultaneously, while having a smaller dynamic range means that a number system will be more prone to overflow and underflow.

1.4 Fixed-Point Arithmetic

The simplest possible quantization of a number space is to "sample" it with even spacing between samples. This is a simple extension of the integers: The integers are the result of sampling the real numbers with a sampling interval of one. In other words, we put all of the bits of our number in front of the decimal point. Computer integers also have a finite number of bits, b, and can only represent numbers between -2^{b-1} and $2^{b-1} - 1$ (for two's complement signed integers). A number line of the integers is in Figure 1.2. Many programmers are intimately familiar with the integers and what you can and can't do with them algebraically. The integers are actually very well-behaved in this sense, with the exception of division.

If we choose a sampling interval of 2^{-k} (typically with $k > 0$), we get fixed-point arithmetic. A fixed-point number line is shown in Figure 1.3 compared

−8	−7	−6	−5	−4	−3	−2	−1	0	1	2	3	4	5	6	7

FIGURE 1.2

A number line of the 4-bit two's complement signed integers.

against a number line of the integers. This scaling factor puts some bits of an integer in front of the decimal point, and some bits behind the decimal point. Fixed point can be thought of as a mapping of integers to non-integer values on the number line. This way, we can represent non-integers with a given granularity. What we gain in precision, however, we lose in range. With a 32-bit integer, the size of the range is about 4 billion, but if we put half of those bits after the decimal place, we can now only represent numbers between −32 768 and 32 767. However, we can now represent 0.5 and 0.25 precisely and approximate 0.1 usefully.

There is a general notation for fixed-point number formats, which is to use a Q when indicating the number of bits in front of and behind the decimal place. A Q16.48 number is a 64-bit number with 16 bits in front of the decimal place and 48 bits after the decimal place, indicating $k = 48$. This number system has the same range as the 16-bit integers, but now has the capability to represent most common decimals with precision.

Fixed-point operations are easy to do when you have an integer computer. A fixed-point operation is computed by combining integer operations with some shifting and bit manipulation. Addition of fixed point numbers just requires that the decimal places in the two numbers be lined up, while multiplication involves a shift of the double-width result of the multiplication. The product of two fixed point numbers includes the product of its bias. The product of two integers of n bits is a $2n$ bit number, and while we usually discard the top of the result for integer calculations, we need to use some bits from it for calculations in fixed point: The product of a pair of Q16.48 numbers is a Q32.96 number, so most of the bits of our result end up in the high part of the product.

Achieving good numerical accuracy on long arithmetic chains in fixed point is a difficult task that involves precise tracking of number ranges, significance, and precision. Fixed point arithmetic chains can also involve temporarily extending the precision of numbers to keep significance while numbers move up

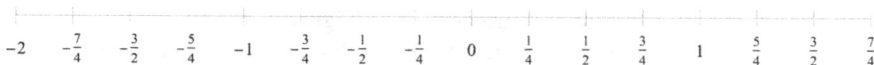

−2	$-\frac{7}{4}$	$-\frac{3}{2}$	$-\frac{5}{4}$	−1	$-\frac{3}{4}$	$-\frac{1}{2}$	$-\frac{1}{4}$	0	$\frac{1}{4}$	$\frac{1}{2}$	$\frac{3}{4}$	1	$\frac{5}{4}$	$\frac{3}{2}$	$\frac{7}{4}$

FIGURE 1.3

A number line of the 4-bit two's complement fixed point numbers with two bits behind the decimal place ($k = -2$ or Q2.2 format).

and down the range of magnitudes. A `Q48.16` fixed point number will represent numbers close to 2^{40} with fantastic precision, but numbers less than 0.01 will be somewhat cramped. This means that keeping precision through an operation chain is best done by making sure that the intermediate results are represented with the largest possible integer values. If you have a number that you are sure will be less than 100, you are losing precision by keeping more than 8 bits ahead of the decimal place. If you want to add that number to a number that is 10 000 or greater, you now have to switch the locations of your decimal place for the final result. Tracking all of this is a difficult task and requires thorough knowledge of the exact operating environment of your arithmetic to navigate the path between imprecision and overflow.

It is also possible to work with a sampling interval that is not a power of two, although this does not fit the strict definition of "fixed point". For example, stock exchanges around the world send prices represented as "integer in units of 10^{-4} dollars" (so a price of 10 000 corresponds to one dollar), which can be thought of as decimal fixed point. This is a good way to ensure that every valid price can be represented exactly, and since the exchange has no need to do arithmetic on prices aside from a few comparisons, they don't mind that this is a coarse-grained representation.

The Nasdaq learned the dynamic range problems of this approach the hard way in May of 2021. Using a price unit of "integer 10^{-4} dollars" and a 32-bit integer, the exchange could only represent a maximum price of \$429 496.72, and the stock of Berkshire Hathaway exceeded this price. For some time, the Nasdaq reported very cheap prices for Berkshire Hathaway due to the integer overflow!

Fixed point gives us a straightforward trade of precision with dynamic range, and if we know the expected range of operations and have a limited computation environment, it can be a useful tool. However, this tool comes at a significant cost in development time and mental effort.

1.5 Decoupling Precision from Magnitude

The idea of floating point is driven by the desire to decouple the magnitude of numbers from their precision. Fixed point numbers are more precise when they are larger, but that is an undesirable property. Scientists tracking significant figures in lab work also face the same dilemma, and their solution is scientific notation:

$$2.48 \times 10^9$$

This number is 2 480 000 000 with three significant digits, indicating that it is only known to within ±5 000 000. The first number in the notation, also called the **significand**, indicates the precise value of the number, while the exponent of the ten indicates its magnitude. The significand has a dynamic range of 10, since it must always have a leading digit before the decimal place, while the exponent determines dynamic range of the format. Similarly, the exponent does not affect the precision of the number format at all.

For numerical analysis, we are less concerned with tracking significance than we are with keeping it, so we will store the significand with as many bits as possible rather than trying to stick to the minimum required size for a number. Since computers work in binary, we will swap over to base two, and we will use enough exponent bits to represent all of the numbers that we think might be reasonable. Only six exponent bits gives us equivalent dynamic range to a 32-bit integer.

By decoupling precision from range, we also simplify our ability to analyze our numbers and our calculations. This trades off against making operations more challenging to compute. It is this idea that brings us to floating point. We store a number as a tuple, explained in the next chapter, representing a significand whose decimal point is allowed to float around based on the value of an exponent.

Check Your Understanding

Problem 1.1. Come up with a formula for the precision of integers relative to their magnitude.

Problem 1.2. Come up with a formula for the precision of rational numbers represented as a pair of integers.

Problem 1.3. Representing rational numbers as pairs of integers, write a function that computes the product of rational numbers without normalizing the fraction. Generate 10 000 random fractions with numerator and denominator in the range $[1, 100]$. What is the average number of multiplications you need to reach overflow? What is the minimum number?

Problem 1.4. Representing rational numbers as pairs of integers, write a function that computes the product of rational numbers and write a normalization function. Generate 10 000 random fractions with numerator and denominator in the range $[1, 100]$. Compute the product of these numbers, normalizing any time you would risk overflow. How many normalization steps do you need? Compared to the product of 10 000 random floating point numbers in the range $[0, 1]$, how much slower is it to work in the rational numbers?

Problem 1.5. Several virtual machines from cryptocurrencies use 32-byte integers and have no floating point. They replace floating point with `Q128.128` fixed point. What is the dynamic range of this number system? Write a multiplication algorithm for this fixed-point number system and compare its speed to double-precision floating point on your computer.

2

Numbers in Floating Point

Floating point stores numbers in scientific notation, as a tuple of three separate numbers, a sign bit, an exponent, and a mantissa field that represents the "tail" of the significand. This is a compact representation that gives precision that is independent of magnitude. However, the devil is in the details. While the normal numbers follow a nice pattern, there are several exceptions that are required to give the floating point numbers nice properties.

Floating point also contains several special numbers, including two zeros, two infinities, and several representations of invalid numbers. This keeps calculations in floating point mathematically closed and gives nice properties for numerical calculations down to 0 and up to infinity, while still maintaining space to represent errors that would otherwise break the number system.

While the design may seem convoluted at points, the design choices that define floating point have been made to reduce the cognitive load on programmers while approximating the real numbers as closely as possible.

2.1 Extending the Real Numbers

We would like to keep calculations in the floating point numbers algebraically closed while still representing a small subset of the real numbers. To do this, we begin by extending the real numbers to have all of the numbers we need. We are going to have zero in our quantized number system, so we are going to need to have some representation of $\frac{1}{0}$ in our number system. This means that we need to find a way to map infinity into our number system.

Mathematically, we can do this by projecting the real numbers onto a circle as shown in Figure 2.1. This figure shows a projection called the affine extended real number line, where we project the real number line onto a circle above the number line. Projection lines from the circle to the number line go from the top of the circle to a point on the number line. This uniquely maps every real number to each point on the circle, and puts positive and negative infinity at the point on top of the circle.

We then have a choice about our infinities and zeros. We can choose to have both infinities, allowing us to represent arbitrarily large positive and negative numbers separately, or we can choose to have a single infinity corresponding

DOI: 10.1201/9781003565543-2

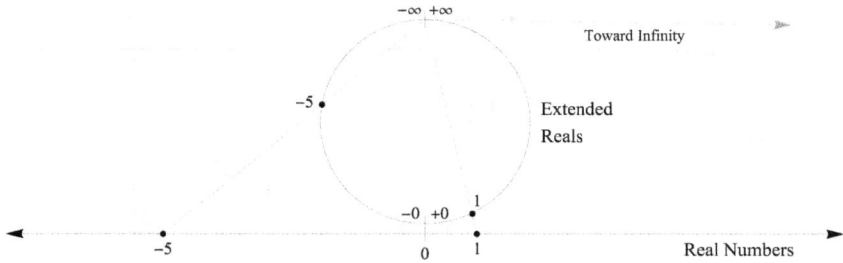

FIGURE 2.1
Projection showing the mapping of the real number line onto the extended real numbers. The extended real numbers are represented by the circle, which has the infinities at its zenith.

to the point on the circle. In floating point, we use two infinities. This allows us to have calculations overflow in both directions while keeping the infinities ordered with respect to the numbers. We know that positive infinity is greater than all the positive numbers, while if we had one infinity, we would not have that guarantee. Corresponding to the two infinities, we have two zeros, one positive zero and one negative zero. That allows us to give $\frac{1}{+0} = +\infty$ and $\frac{1}{-0} = -\infty$, keeping the direction from which the denominator approached zero.

What remains is for us to choose points around the circle to use for our quantization. We do that in floating point by unrolling the circle into a line and then choosing points that are distributed to decouple precision and accuracy using a binary scientific notation. Our encoding will also need to include several special numbers to keep the algebra system closed, like the pair of infinities and their corresponding zeros.

2.2   Binary Scientific Notation

An integer in binary looks like:

$$25\,049\,568 = 1\,0111\,1110\,0011\,1001\,1110\,0000_b$$

Note the b subscript, indicating a binary number. For scientific notation, we would remove powers of ten from the number and move them to an exponent, while for binary scientific notation we will remove powers of two, making the process simpler for computers. Translating into binary scientific notation, we get:

$$25\,049\,568 = 1.4930706024169922 \times 2^{24}$$
$$= 1.01111111000111100111100000_b \times 2^{11000_b}$$

Since multiplying a number by 2^k is equivalent to shifting that number to the left by k bits, computers and circuits can work in this form efficiently. Every number in scientific notation has a single digit in front of the decimal place. In binary, that digit can only have one possible value: It must be one. This means that we do not need to store the one, since we know it's there, and it will take up valuable encoding space that we can otherwise use to extend precision or dynamic range. Binary scientific notation combined with the bit-saving trick of implied bits motivates the design of the floating point numbers.

A floating-point number is a 3-element tuple containing:

1. A sign bit.

2. An exponent with length *eBits*, stored in an offset-binary format.

3. A mantissa of length *mBits*, representing the fractional bits of the significand.

Our floating point number (usually) maps to the real number corresponding to:

$$F = (-1)^s 2^e (1.m)$$

where s is the sign bit, e is the exponent (not the mathematical constant), and m is the mantissa. The sign bit is the simplest field, with a zero for a positive number or a one for a negative number. The exponent is an integer indicating the power of two to multiply by the significand. This exponent is not a normal signed integer, however. It is stored as an unsigned number with a negative bias (*eBias*). This way, small numbers in the exponent field correspond to floating point numbers that are close to zero, while large numbers correspond to large numbers, so you can compare floating point numbers by comparing them as though they were integers. The final field, the mantissa, shows the tail of the significand after the decimal place. This mapping works for the floating point numbers called the **normal** numbers, but note that there is no way to represent zero as a normal number.

A few examples of normal numbers are:

$$1 = 1.0_b \times 2^0 \longrightarrow (s = 0, e = eBias, m = .0)$$
$$-1\,337 = -1.0100111001_b \times 2^{10} \longrightarrow (s = 1, e = eBias + 10, m = .0100111001)$$
$$\tfrac{5}{4096} = 1.01_b \times 2^{-10} \longrightarrow (s = 0, e = eBias - 10, m = .01)$$

To save space, we have dropped the trailing zero bits of the mantissas here and kept only the leading bits close to the decimal point. This is the opposite of how integers are usually notated, where leading zeros are dropped. Numbers with an infinite decimal are also approximated by normal numbers as long as they are within the range of available exponents by rounding the mantissa to length $mBits$. The number 0.1 is a repeating decimal in binary:

$$0.1 = 1.\overline{1001}_b \times 2^{-4} \approx 1.1001100110011001100101_b \times 2^{-4}$$
$$\longrightarrow (s = 0, e = -eBias - 4, m = .1001100110011001100101_b)$$

Similarly, we can round irrational numbers to the nearest normal:

$$\pi \approx 3.1415927410125732421875 = 1.1001001000011111101101_b \times 2^1$$
$$\longrightarrow (s = 0, e = 1 - eBias, m = .1001001000011111101101_b)$$

The normal numbers all have a precision equal to $mBits + 1$ bits, and a dynamic range of $2^{eBits-2}$. As we will see, this is a small compromise on precision compared to the largest integers and high-precision fixed-point formats, but it comes with the benefit of gigantic dynamic range and normalized precision across that range. We get that dynamic range and precision by having a very unequal number density of numbers across the number line. Density of numbers increases exponentially as magnitude decreases, with half of the numbers in the format being in the range $[-1, 1]$. A density plot for a floating point format is shown in Figure 2.2.

The exponent also denotes whether a number is normal or not. An exponent of all zeros (the smallest exponent) or all ones (the biggest exponent) indicates a special number, while exponents between $000\ldots01$ and $111\ldots10$ are the normal numbers. These interpretations are shown in Table 2.1. The interpretation of the mantissa changes based on the regime, although the sign bit preserves its function in all but one case. The non-normal regimes are:

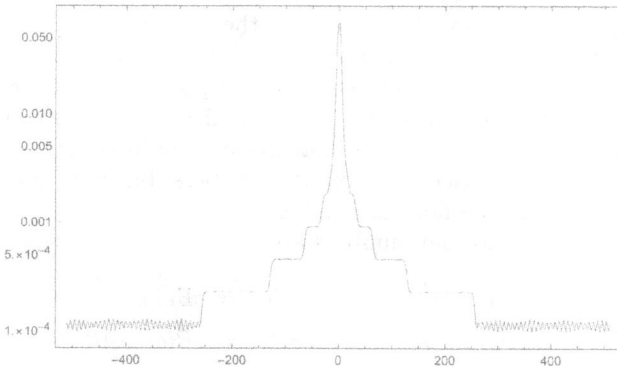

FIGURE 2.2
Log plot of a smoothed density function representing the distribution of numbers in a floating-point format with $eBits = 4$, $eBias = -7$, and $mBits = 3$. Ripples at magnitudes greater than 256 are an artifact of smoothing.

TABLE 2.1
The regimes of floating-point encoding of real numbers.

Exponent	Mantissa	Regime	Value
0	$m = 0$	Zeros	± 0 depending on sign bit
	$m \neq 0$	Subnormals	$(-1)^s 2^{eBias+1}(0.m)$
$2^{eBits} - 1$	$m = 0$	Infinities	$\pm\infty$ depending on sign bit
	$m \neq 0$	NaNs	"Not a number"
All others	All	Normals	$(-1)^s 2^e(1.m)$

- The zeros, ± 0, which indicate a zero or infinitesimal value.

- The subnormal numbers, which allow the floating point numbers to smoothly go down toward zero.

- The infinities, $\pm\infty$, which represent numbers that are too large to represent.

- The NaNs, which indicate that the floating point datum is actually not a number.

The construction of the significand of the number from the binary mantissa depends on the exponent to work appropriately in each regime. If the exponent is zero, the significand is $0.m$, with an implied zero in front of the mantissa. These numbers are the zeros ($m = 0$) and the subnormal ($m \neq 0$) numbers. If the exponent is all ones, and the mantissa is zero, we have an infinity, and all other mantissas indicate "Not a Number" and can carry some other payload in the mantissa. Neither case has a significand.

A number line of a hypothetical floating point format with exponent range $[-2, 2]$ and one mantissa bit is shown in Figure 2.3, combining all five of these regimes. We see the exponential density of the normal numbers in the range $[-6, 6]$, and the number line is bounded by two infinities. As we get toward zero, the transition to the two zeros is smoothed out with subnormal numbers on each side of zero, with magnitudes less than 2^{-2}.

FIGURE 2.3
Number line for a reduced-precision version of floating point. This format has an exponent range of $[-2, 2]$, and only one mantissa bit. The numbers are exponentially distributed, and we have two infinities and two zeros.

2.3 Floating Point Bit Layouts

The tuple of three parts that makes up a floating point number is packed into a small structure of bits, usually 32 (for single-precision) or 64 bits (for double-precision). These two formats are the main formats supported by processors for numerical calculations, and are often called `float` and `double` in programming languages. Together with two other binary formats with less use, the half-precision 16-bit format and the quad-precision 128 bit format, these make up the "mainline" floating point formats. These formats balance precision and dynamic range at each size, and are full-featured floating point formats. These formats, in the floating point standard, are referred to by the names `binary16`, `binary32`, `binary64`, and `binary128`, but we will use the common names going forward.

The elements of the floating point number are packed into words with the sign bit as the most significant bit, the exponent next, and the mantissa in the remaining least significant bits. The formats and their bit widths are summarized in Table 2.2, and shown in Figure 2.4. These are then stored as a single word of the appropriate width on the machine, meaning that the exact byte layout in memory depends on the way the machine stores words. As the total width of the format grows, the number of exponent bits grows more slowly than the number of mantissa bits, although both precision and dynamic range grow equally.

As with the integers, larger floating-point formats are more difficult to operate on. On current CPUs, operations on single-precision floating-point numbers are about equal in speed to operations on 64-bit integers, while operations on double-precision floats are slightly slower. Quad precision floating-point numbers do not have hardware support, so operations are very slow [3].

The numerical properties of each format are shown in Table 2.3. The half precision floats cannot represent numbers larger than $65\,504$ ($1111\,1111\,1110\,0000_b$), which is a relatively "human" magnitude. The single

TABLE 2.2
Binary properties of the four mainline binary floating point formats from half precision to quad precision. Single and double precision are the two main formats. All formats have one sign bit.

Size Name	Width	Exponent Bits	Bias	Mantissa Bits
Half (FP16, `H`)	16	5	-15	10
Single (FP32, `F`)	32	8	-127	23
Double (FP64, `D`)	64	11	$-1\,023$	52
Quad (FP128, `Q`)	128	15	$-16\,383$	112

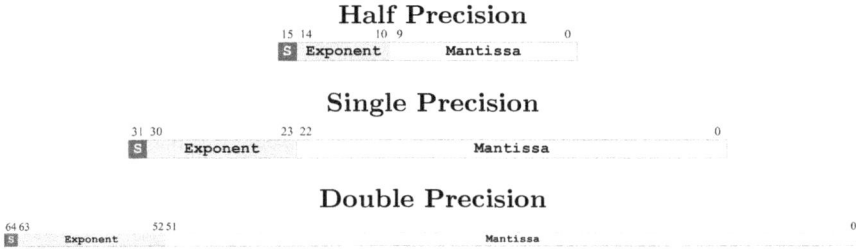

Half Precision

Single Precision

Double Precision

FIGURE 2.4
Binary layout of three floating-point types within their respective binary words.

precision floats have enough dynamic range to represent the size of the known universe in nanometers as well as the radius of an electron in light years, although the available precision is relatively limited, and both quantities will be rounded to a value that is known to be wrong. By the time we reach double precision, the number range that can be represented is wide enough to include quantities like "googol" (albeit not exactly), and there is enough precision available to represent most physical quantities to within their known error bounds. However, as we will see, excess precision is good for avoiding inaccuracy in long computing chains, and excess dynamic range prevents overflows and underflows.

Additionally, the precision level tells us the range of integers that can be represented exactly. The single-precision floats can represent 24-bit integers exactly, but not 25-bit integers. The doubles can represent any integer up to 53 bits exactly. This is why some programming languages can get away with not having an integer type: If you restrict your integer range a little bit, the functionality of the floating-point numbers is a strict superset of the functionality of the integers.

Going forward in this book, we will be using a few representations to notate numbers in a given binary format. A number in floating point will be shown

TABLE 2.3
Numerical properties of the four mainline binary floating point formats from half precision to quad precision.

Name	Precision (bits)	Dynamic Range	Smallest Normal	Largest Normal
Half	11	2^{30}	0.000061	65 504
Single	24	2^{254}	1.18×10^{-38}	3.40×10^{38}
Double	53	2^{2046}	2.23×10^{-308}	1.80×10^{308}
Quad	112	2^{32766}	3.36×10^{-4932}	1.19×10^{4932}

with a subscript indicating its type, with H for half precision, F for single precision (many languages call these numbers `float`, so we will abbreviate with F), and D for double precision. Examples of this representation are:

$$64_F \qquad\qquad -56_D \qquad\qquad 0.125_H$$
64 in single precision | −56 in double precision | 0.125 in half precision

There will be times when it is better to separate the significand and exponent. When this happens, we will use a notation similar to the "e" notation that is used for scientific notation:

$$\pm[\text{Significand}]\,[\text{\tt Format}]\,[\text{Exponent}]$$

As with integers, a $_b$ subscript indicates that any of these numbers is in binary rather than in decimal. The three examples from above will look like:

$$64_F \qquad\qquad -56_D \qquad\qquad 0.125_H$$
$$1.0 \times 2^6 \qquad 1.75 \times 2^5 \qquad 1.0 \times 2^{-3}$$
1.0F6 | $[-1.11_b\text{D5}]$ or $[-1.75\text{D5}]$ | 1.0H(−3)

Although this scientific notation is more verbose than the number with a subscript, some corner cases and pathological examples benefit from looking at them in this way. We also include the implied leading bit of the significand in the notation even though it is not in the stored number.

Given a set of 32 bits, a mapping of that number's integer value to its floating point value is shown in Figure 2.5. For each of the positive and negative numbers, this mapping is monotonic, and roughly traces out a linearized exponential function. Comparing floating point numbers can be mostly done by comparing the numbers as integers. Creating this monotonic mapping is part of why offset-binary format is used for the exponent rather than two's complement.

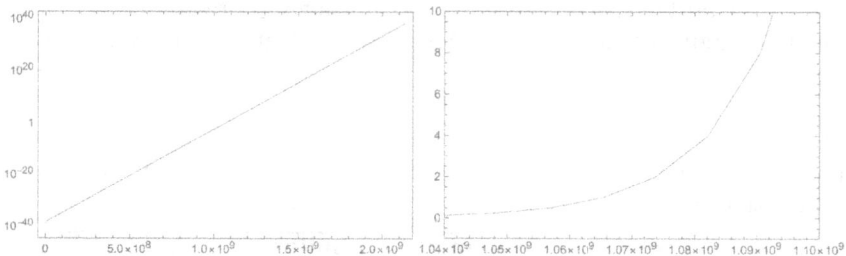

FIGURE 2.5
Left: Log plot mapping the integer value of a word when read as a two's complement integer (x axis) to the floating point value of the same word (y axis), showing the positive half of the range. Right: Zoomed in linear plot showing the shape of the curve.

Measuring Format-Aware Error

Since our precision is measured in bits and, for the normal numbers, is fully decoupled from dynamic range, it helps to track precision loss during operations in **units in the last place** (**ULPs**). The size of one ULP for a number of a given exponent, e, is:

$$2^{e-mBits}$$

Many operations have up to half a ULP of error corresponding to rounding an inexact result, but some operations are worse. This gives us a measure of error that directly translates to the layout of the format and is decoupled from the value of the operands. This measure of error allows us to look at the error of calculation chains in a range-agnostic way when measuring how accurate they will be.

2.4 Zeros and Infinities

Floating point has two zeros and two infinities. The two zeros are one of the more confusing parts of floating point, but they go hand-in-hand with two infinities. The infinities are the only floating point numbers that are not real numbers: Infinity is not a real number since the real numbers can get arbitrarily large, but it is a floating-point number. In the real numbers, functions can approach infinity as you move smoothly toward some value a, but will keep growing and growing until being undefined exactly at a. While a might be a floating point number, there is no way to approach a smoothly in the quantized space of floating point, and leaving a hole in the value of a function at a is a dramatic thing to do. Thus, instead of infinity being something that you move toward, floating point makes the infinities valid numbers.

When you have negative and positive infinities, you need to be able to know what $\frac{1}{\infty}$ and $\frac{1}{-\infty}$ are. That gives us two zeros. Several functions approach zero from the negative values at a given point, and if you don't have a negative zero, the inverse of those functions will go to positive infinity instead of negative infinity as expected. A case of this is the simple function:

$$f(x) = \frac{1}{10^{-10}x}$$

With a negative zero, when x gets close to 0, the denominator calculation will underflow to negative zero, and $f(x)$ will be negative infinity, as expected. Without a negative zero, the denominator will underflow to zero, and $\frac{1}{0}$ could evaluate to positive infinity, which is very far from the correct answer! In reality, what would happen is that infinity would have to become unsigned, representing either positive or negative infinity. Having the two zeros allows $\frac{1}{x}$ to have a determinate value for all possible values of x.

One misconception about the two zeros is that they represent infinitesimal values, where negative zero is a tiny negative number and positive zero is a tiny positive number. While tiny negative and positive numbers can become their respective zeros, it's not accurate to think of them that way. The zeros are zero, but the sign indicates the direction from which we approached zero. "Zero approached from the left" is negative zero, while "zero approached from the right" is positive zero. The zeros can be created by underflow, in which case they represent an infinitesimal value, but both zeros can also arise exact calculation.

Positive zero is the "default" zero when it is indeterminate which zero to use, but almost all cases have a determinate sign for zero. If we modify our example from before:

$$f(x) = \frac{1}{10^{-10}x - 1}$$

When $(10^{-10}x)$ yields 1, even if it does so inexactly, $f(x)$ will evaluate to positive infinity.

Similarly, the infinities are also often thought of as being "too big to represent", which is sometimes true, but infinity also often represents a value that is actually infinite. There are two types of operations that can yield infinities when done on normal numbers, overflows and divisions by zero. In the former case, the infinity is a number that is too big to represent, but in the latter case, the infinity is actually infinity.

2.5   Subnormal Numbers

Looking at our formula for the value of the normal numbers,

$$(-1)^s 2^e (1.m)$$

the smallest positive normal number has the minimum normal exponent and a mantissa of zero:

$$2^{eBias+1}$$

This is a small number, and for the single-precision floats it is 2^{-126}. The distance between this number and the next number above it is one ULP away, while the next number below it might be zero. That ULP, for the single-precision floats of magnitude 2^{-126} is:

$$\mathrm{ULP}\left(2^{-126}\right) = 2^{-126-23} = 2^{-149}$$

The size of one ULP is a lot smaller than the distance between 2^{-126} and zero. The subnormal numbers, also called "denormalized numbers" or "denorms", cover the gap between the normal numbers and zero by continuing toward

FIGURE 2.6
Comparison of a floating point number line with subnormal numbers (top) against a number line with an extension of the normal numbers (bottom). With the normal numbers only, there is a big jump to zero, while the subnormal numbers create a gradual transition.

zero gradually. Figure 2.6 shows why the subnormal numbers are valuable by comparing them to using that encoding space for additional normal numbers. The use of subnormal numbers costs a small amount of precision in a small numeric range, but avoids a large loss of precision near zero. They allow precision to degrade slowly rather than all at once.

Even though the subnormal numbers have an exponent of 0 in the exponent field, they still use the exponent of $eBias + 1$. To approach zero gradually, the subnormal numbers have an implied leading zero instead of an implied leading one. This leads to the formula for the value of a subnormal number:

$$(-1)^s 2^{eBias+1}(0.m)$$

This allows us to walk down toward zero with a monotonically decreasing step size rather than having a single big jump at the end.

The danger of the subnormal numbers is that the implied leading zero means that they have degraded precision. The precision of the subnormal numbers is like the precision of the integers, and is dependent on magnitude. Operations on subnormal numbers will still accumulate error in ULPs, so they cause relative error to accumulate faster than operations on the normal numbers. They are still a lot more precise for small numbers than rounding to zero or to a much larger normal number [4].

Operations on subnormal numbers used to need help from microcode, and were significantly slower than operations on the normal numbers, but processors have started to integrate floating point hardware that works equally fast on subnormal numbers as on normalized numbers [5]. For performance on these platforms, it is usually possible to turn off support for subnormal numbers, and just flush them to zero instead [3].

2.6 NaNs (Not-a-Number)

The last parts of the floating-point encoding space are the "Not-a-Numbers" (NaNs). As the name suggests, these are not numbers. There is also a large amount of encoding space devoted to them, since every floating-point value with an exponent of all ones and a non-zero mantissa field is a NaN. A single-precision floating-point NaN breaks down into the parts:

$$\text{NaN}_F = \quad \begin{array}{ccc} \text{S} & \text{Exp} & \text{Mantissa} \\ \text{X} & \text{11111111} & \text{XXXXXXXXXXXXXXXXXXXXXXX} \\ \text{Any} & e\mathit{Max} & \text{Any nonzero} \end{array}$$

There are two types of NaNs: Quiet NaNs (qNaN) and signaling NaNs (sNaN). Quiet NaNs pass through operations, while signaling NaNs raise an exception (specifically the invalid operation exception). Once an exception is raised from a signaling NaN, the NaN is "quieted" and turned into a quiet NaN. The distinction between quiet and signaling NaNs is the top bit of the mantissa. A quiet NaN has a leading one bit, while a signaling NaN has a leading zero bit:

$$\text{qNaN}_F = \quad \begin{array}{ccc} \text{S} & \text{Exp} & \text{Mantissa} \\ \text{X} & \text{11111111} & \text{1XXXXXXXXXXXXXXXXXXXXXX} \\ \text{Any} & e\mathit{Max} & \text{Any with leading 1} \end{array}$$

$$\text{sNaN}_F = \quad \begin{array}{ccc} \text{S} & \text{Exp} & \text{Mantissa} \\ \text{X} & \text{11111111} & \text{0XXXXXXXXXXXXXXXXXXXXXX} \\ \text{Any} & e\mathit{Max} & \text{Nonzero with leading 0} \end{array}$$

The general idea of signaling NaNs is that the developer can set up a signaling NaN at the point of an error, and use floating point exceptions to catch the point of the error.

The rest of the mantissa field is free to carry any payload. There are several different proposed uses for these payload bits for carrying information about errors and other similar uses, but none are standardized or uniformly implemented across platforms. An important use of these bits is NaN boxing, where a separate type is carried inside the payload bits of a NaN, allowing a union of a floating point number and another type to share the same bits with efficient type detection (see Section 8.4) [6].

Operations that have no determined value, such as $\frac{0}{0}$, produce a NaN to indicate an error in the arithmetic that results in an indeterminate answer. Additionally, operations that involve a NaN will propagate a NaN to the result because any operation with a NaN input cannot produce a determinate result. The exact NaN generated depends on the behavior of the hardware. Some processors will propagate one of the input NaNs to the output, and many will produce a new NaN, either a specific "default" NaN or one with scrambled payload bits [7].

Beyond the level of the processor, some programming languages also specify a canonical NaN and additional rules on production and consumption of NaNs. Incoming NaNs to these environments get canonicalized, turning them into the one canonical NaN, and any calculation that produces a NaN has an additional check to canonicalize its output. Additionally, several languages, including JavaScript [8], do not support signaling NaNs. These math libraries will usually treat signaling NaNs as quiet NaNs, but some will have errors or bugs (possibly even critical security exploits) if you find a way to get a non-standard NaN into a program.

2.7 Other Floating-Point Formats

There are two other classes of floating-point formats that are different from the main binary formats but deserve a mention. These should generally be considered special-purpose numerical formats:

- Reduced-precision floating-point formats

- Decimal floating point

Reduced-precision formats are primarily for machine learning models, and the smallest formats specifically only work well for large models. Decimal floating point is used in some niches for accounting calculations.

Reduced-precision formats have gained prominence recently due to their use in large language models and other large-model machine learning systems. The primary benefit of these formats is that they can compress the storage of model weights while maintaining the dynamic range of larger integers. The main reduced-precision formats are shown in Table 2.4. These formats largely come from CPU and GPU vendors, as well as the Open Compute Project, which is a consortium of vendors. Bfloat16 and E5M2 have similar NaNs and infinities to mainline floating point, but the smaller formats have no infinity and have normal numbers all the way up to the max exponent. E4M3 compromises, placing two NaNs with exponent and mantissa fields of all ones,

TABLE 2.4
Binary properties of proposed reduced-precision floating-point formats, including the number of zeros, infinities, and NaNs in the encoding space. All formats have one sign bit. Names indicate the number of bits in the format, with "E" and "M" labeling the number of exponent and mantissa bits.

Name	Width	Exponent Bits	Bias	Mantissa Bits	Zeros/ Infinities	NaNs
Bfloat16	16	8	-127	7	2 / 2	254
FP8-E5M2	8	5	-15	2	2 / 2	6
FP8-E4M3	8	4	-7	3	2 / 0	2
FP4-E3M0	4	3	-3	0	0 / 0	0
FP4-E2M1	4	2	-1	1	2 / 0	0

although the remaining numbers with an exponent of all ones are normal numbers. All formats except E3M0, which is a hypothetical format with no zeros, still have two zeros and the standard subnormals [9].

Bfloat16 and E5M2 result from the truncation of common floating-point formats to a smaller binary size. Bfloat16 is a truncated version of single-precision float from 32 bits to 16, and E5M2 is a truncated version of half-precision float from 16 bits to eight.

Some of these formats are not used for calculations, and the ones that are used for calculations are often used in matrix operations that accumulate to a larger-precision format. For example, several large language models will use matrix operations with FP8 inputs that accumulate results in half precision. Outside of the matrix multiplications, half-precision floating point is still common. The smallest formats are often only used for storage, and calculations happen in a wider, more precise numeric format.

Decimal floating point is an IEEE-754 standard that is primarily used for calculations involving money. Decimal floating point does not use 2 as its exponent radix, but 10. This introduces a number of problems for the format that make things more difficult than in the binary formats. There are three standardized decimal floating-point formats, a 32-bit format, a 64-bit format, and a 128-bit format. These formats are shown in Table 2.5.

The encoding of numbers in decimal floating point is more complex than binary floating point, and there are two encodings. Both encodings replace the exponent with a "combination" field that encodes the exponent and the high bits of the significand. Although the exponent range is smaller for decimal floating point than the binary format equivalent, the full combination field is several bits larger. In the first encoding, the significand (S) is encoded as a binary number with the high bits covered by the combination field and the remaining bits in the mantissa:

$$(-1)^s 10^e S$$

In the second encoding, the mantissa is a densely packed decimal format using an optimized encoding for packing binary-coded decimal digits together. This encoding packs three decimal digits into ten binary bits for maximum information density, and then uses packs of 10-bit pieces to create the mantissa.

TABLE 2.5
Decimal properties of the decimal floating-point formats.

Name	Width	Exponent (10^e) Min	Max	Significand Digits
Decimal32	32	−95	96	7
Decimal64	64	−383	384	16
Decimal128	128	−6 143	6 144	34

Broadly speaking, the use of decimal numbers means applying some forms of compression in order to take advantage of the binary encoding space available.

Unlike binary floating point, both encodings have multiple forms of the same number, so decimal floating point calculations sometimes need to regularize the value of their results. Otherwise, the operations on floating-point numbers translate to decimal floating point. The decimal floating point numbers can represent powers of 10 exactly, so while binary floating point formats must approximate the number 0.1, decimal floating point has an exact representation.

Despite being a standard format, hardware support for decimal floating point is relatively limited, with IBM offering support in its CPUs and mainframes [10], but no support from either the x86 or Arm architectures. However, several software libraries provide support for decimal floating point if it is of interest.

Check Your Understanding

Problem 2.1. Write a program that constructs a single-precision floating-point number from a tuple of integers. Use that program to construct the following floating point numbers:
- 1.0
- 47 919 936.5
- 0.013843536376953125
- $-\infty$
- 2^{-140}

Problem 2.2. Write a program that breaks down a double-precision floating-point number into a tuple of integers. Use that program to construct the following sets of integers:
- Sign: 0, Exponent: 1 300, Mantissa: 0
- Sign: 1, Exponent: 1 023, Mantissa: 100 000 000 000
- Sign: 0, Exponent: 1 000, Mantissa: 2^{51}
- Sign: 0, Exponent: 0, Mantissa: 123 456 789
- Sign: 1, Exponent: 96, Mantissa: 987 654 321

Problem 2.3. How many possible NaNs are there in double-precision floating point?

Problem 2.4. Of all possible single-precision floating-point numbers, what fraction are greater than 1 337? What fraction of half-precision floating-point numbers are greater than 1 337?

Problem 2.5. Using a random integer generation algorithm (e.g., C rand()), write a function that produces a random single-precision floating point number drawn uniformly from the range $[1, 2]$. Subtract 1 from this result to produce a random floating point number from the range $[0, 1]$. Are there any floating point numbers in the range $[0, 1]$ that your function will never produce?

Problem 2.6. Write a function that takes the logical AND of two positive integers stored in double-precision floating point. Ensure that your function works for all positive integers that can be represented in double precision (including those with exponents over $1\,000$). Does this function always calculate the exact AND of the two numbers?

3

Sources of Error

Floating point operations can introduce error through rounding and through exceptional conditions. Rounding and the exceptions are notorious for creating calculation problems, partly due to using a number format designed for numerical analysis to do generalist arithmetic.

Rounding is one of the most potent aspects of floating point, both in terms of its power and the challenges involved in using it. There are four rounding modes that are available to be used, and any time an inexact calculation occurs, the result is rounded. Rounding is often responsible for the accumulation of error in calculations, but it keeps your results as accurate as possible while staying in a fixed bit width.

Similarly, there are five exceptions in floating point, from division by zero to an inexact exception, which are signaled to the user to indicate any condition where a calculation does not exactly match what should happen with arithmetic in the real numbers. Exceptional conditions can be triggered by all floating-point operations. If an operation does not hit an exception (and therefore does not round), it has not added any error to the result of that calculation.

3.1 Rounding and Rounding Modes

One part of floating point that cannot be ignored or avoided is rounding. Rounding is an important feature of any numeric format of finite length. Integer calculations round by truncation. If you have any fractional bits, they get flushed to zero. Floating point takes rounding more seriously to avoid adding error where possible. There are four default rounding modes for binary floating point that can apply to every operation:

1. **Round to nearest, with ties to even (RTN).** In this mode, rounding goes to the nearest number, but results exactly halfway between two floating-point numbers are rounded toward the number with the even mantissa (the number whose mantissa has a trailing zero). This is the default rounding mode.

DOI: 10.1201/9781003565543-3

2. **Round toward zero (RZ).** In this mode, results are always rounded toward the smaller (in magnitude) of the two options. This mode is equivalent to rounding by truncation, and is usually what people mean when the think of "rounding down".

3. **Round toward positive (round up or RU).** In this mode, results are always rounded toward the higher of the two options.

4. **Round toward negative (round down or RD).** In this mode, results are always rounded toward the lower of the two options.

The last three rounding modes are referred to as **directed rounding** modes, because you control the direction of the rounding, while the first one is often abbreviated as just **round to nearest**. The three directed rounding modes allow you to always round toward the smaller number, rounding up below zero and rounding down above 0, or to always round up or down. A few visual examples of these rounding modes are shown in Figure 3.1.

The round-to-nearest mode is the most accurate of the rounding modes. The behavior of ties going to the even number is unintuitive, but it means going to the "rounder" of the two options, and also balances error in long chains of operations. As an example, three examples from the figure show ties

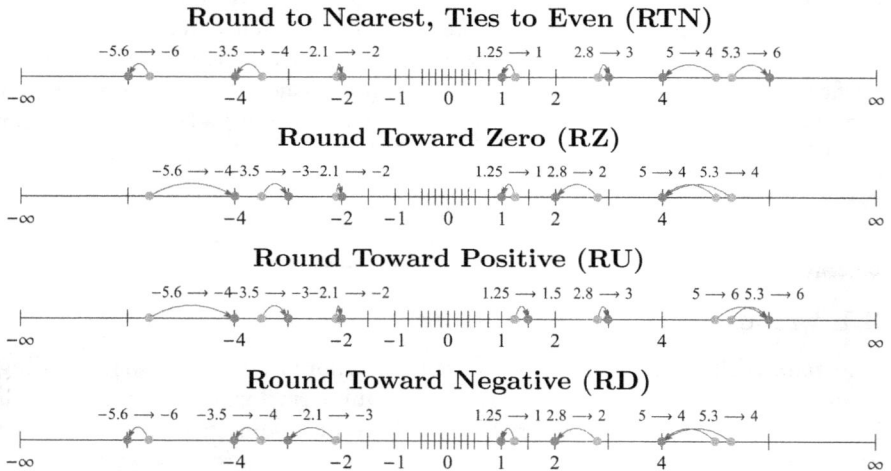

FIGURE 3.1
Examples of rounding of operation results under each of the rounding modes using an example floating point system with 1-bit mantissa and exponent range $[-2, 2]$.

going to the number with an even mantissa:

$$
\begin{array}{rcll}
-3.5 & \to & -4 & \text{instead of } -3 \\
1.25 & \to & 1 & \text{instead of } 1.5 \\
5 & \to & 4 & \text{instead of } 6
\end{array}
$$

Examples which are not exactly between the two options, including -5.6 and 5.3, are rounded as normal to the nearest result. We can also see that the directed rounding modes can cause results to move significantly further than round to nearest, with certain directed rounding modes taking larger steps than might be expected:

$$
\begin{array}{rcll}
2.8 & \to & 2 & \text{instead of 3 for RD and RZ} \\
-2.1 & \to & -2.5 & \text{instead of } -2 \text{ for RD} \\
-5.6 & \to & -4 & \text{instead of } -6 \text{ for RU and RZ}
\end{array}
$$

The classic schoolbook rounding method, shared by children and accountants alike, is not an option in binary floating point. This rounding mode would be called "round to nearest, ties away from zero" since the classic idea of rounding is to choose the larger of the two options when you are halfway between two choices (e.g., 12.5 rounds up to 13). This rounding method tends to accumulate error faster in long operation chains because slightly more than half of the numbers between each floating-point number will round up. With ties to even, the number line alternates between having a bias toward rounding up and a bias toward rounding down, balancing out any rounding bias. Incidentally, this mode is standard in decimal floating point to accommodate accounting calculations, but not binary floating point.

The classic case for rounding to even is the following pathological example. Given two numbers x and y, if you take the sequence:

$$
\begin{aligned}
x_1 &= (x + y) - y \\
x_2 &= (x_1 + y) - y \\
&\;\;\vdots \\
x_n &= (x_{n-1} + y) - y
\end{aligned}
$$

The resulting x_n under round-to-nearest-ties-to-even is either equal to x or x_1 for all $n \geq 1$. Under schoolbook rounding, however, x_n will blow up toward an infinity if the operations used are inexact. For example, let's consider rounding to two decimal places for each operation, with $x = 10$ and $y = 5.5$. The operation chains look like [11]:

Ties to Even	Ties Away from Zero
$x_1 = (10 + 5.5) - 5.5$	$x_1 = (10 + 5.5) - 5.5$
$x_1 = (16) - 5.5$	$x_1 = (16) - 5.5$
$x_1 = 10$	$x_1 = 11$
$x_2 = (10 + 5.5) - 5.5$	$x_2 = (11 + 5.5) - 5.5$
$x_2 = (16) - 5.5$	$x_2 = (17) - 5.5$
$x_2 = 10$	$x_2 = 12$
$\vdots$	$\vdots$

For ties to even, we stay with $x_n = 10$ for all n, while schoolbook rounding will continue until $x_n = 95$. At that iteration, $95 + 5.5$ will round to 100, since we are rounding to two decimal places. Subtracting 5.5 again will return $x_{n+1} = 95$. With ties to even, it is much harder to construct cases where rounding alone creates numerical instability like this. In 2023, this rounding mode cost the author \$2 in extra taxes compared to ties-to-even.

3.2 Errors and Exceptions

Floating point leaves the handling of exceptional conditions up to the programming language, the CPU architecture, and the user. The only official constructs related to error conditions in floating point are the five exceptions. These exceptions indicate that something unexpected or otherwise imperfect has occurred. A floating-point operation that gives no exceptions behaves the same way that the operation would under arithmetic in the real numbers.

Since each of these exceptions is important when thinking about floating point, we will be noting them with a two-letter code for easy reference in mathematical expressions. The five exceptions are:

- Invalid Operation (IO) exceptions occur when an expression has an indefinite result.

- Division by Zero (DZ) exceptions occur when an operation on finite inputs has an exact infinite result.

- Underflow (UF) exceptions occur when an operation's result is too small to be represented.

- Overflow (OF) exceptions occur when an operation's result is too big to be represented.

- Inexact (IX) exceptions occur whenever an operation's result is not exactly equal to its real-valued equivalent.

Not all of these exceptions are errors. Inexact operations, in particular, are commonplace in most numerical code, and generally pass silently.

These codes are the same as the codes used by floating point status flags in some CPUs, and will also be used to notate exceptions that occur in floating-point operations in this book. The notation for operations that we use going forward will look like:

$$[Operation] \overset{\text{CODES}}{\Rightarrow} [Result]$$

Since many of these operations are irreversible, especially operations with an exception, we are also replacing the usual equals sign that appears in math notation with an arrow indicating the production of a result from the operation when we consider floating-point operations. Exception codes will be placed above the production arrow to indicate the exceptions that occur during that operation.

3.3   The Inexact (IX) Exception

The inexact (IX) exception is surprising for people who are users of floating point and are comfortable with inexact calculations—your hardware warns you when that occurs. Inexact exceptions arise any time an operation produces an inexact result. It is both the most common exception and the most commonly ignored exception. More rigorously, the inexact exception indicates that rounding occurred and that the rounded result is not equal to the exact result as computed with infinite precision. It is a normal part of working in floating point, but can still be helpful to know about.

Every inexact operation represents a small precision loss compared to a real-valued calculation, and in cases where you need to track precision loss very closely, the inexact exception is how to handle this. An operation with an inexact exception may add up to one ULP of error (depending on rounding mode) to a calculation, while an exact operation does not add error.

In geometric and numerical calculations, inexact exceptions are expected and usually ignored. However, if you are interested in tracking specific points that create error, it is possible to pay attention to points of inexactness.

3.4   Invalid Operations (IO) and Division by Zero (DZ)

The first two exceptions we will discuss relate to conditions where fundamentally bad inputs are fed to an operation. An invalid operation exception (which we will abbreviate as IO) indicates when an operation is done on inputs that would produce an indefinite result. All operations that signal an invalid operation exception produce a NaN. An example of this kind of operation is:

$$\infty_\mathsf{F} - \infty_\mathsf{F} \Rightarrow ?$$

Depending on the nature of the infinities, the correct result of this calculation could be any number, including ∞, depending on how each of the operands was reached. Since we have none of that history in our floating-point subtraction, we do not try. Instead, floating point signals an invalid operation:

$$\infty_\mathsf{F} - \infty_\mathsf{F} \overset{\mathtt{IO}}{\Rightarrow} \mathsf{NaN_F}$$

However, not all operations that produce a NaN will signal an invalid exception. Any operation on a quiet NaN will output a quiet NaN without this exception, and operations on a signaling NaN will output qNaN while signaling an invalid operation exception. In other words, whenever there is a signaling NaN, that signal will come in the form of an invalid operation exception. Invalid operations can also occur on conversions from floating point to integer if the result would be an integer version of infinity (e.g., a number greater than INT_MAX and a rounding mode that would require that number to be rounded up), and in the case of a signaling comparison involving a NaN. In these operations, they will not produce a result since there is no "NaN" for integers or booleans.

Invalid operation exceptions never occur when all the inputs to an operation are normal numbers or zeros. Usually, you need to have an infinity or a signaling NaN to produce an invalid operation exception. Aside from signaling NaN inputs, which always produce an invalid operation exception, the list of cases in which an invalid operation exception occurs can fit in a few short tables. For reference, the operations producing a NaN and an IO exception are in Table 3.1.

Division by Zero (DZ) is a special kind of operand-related condition which occurs when a floating point operation on a finite input produces an exact infinite result. Division by zero is the obvious case: integer division by zero is a problem, but since we have an ∞ in our number system, we can represent $\frac{1}{0}$ exactly. Any number divided by either zero produces an infinity with an appropriate sign, and signals a division by zero exception. Many systems and programming languages, including Python, will treat a floating-point division by zero as a language-level error condition, but it is not an issue within the floating point number system as long as you can handle the infinities produced

TABLE 3.1
Enumerated list of cases that produce an invalid operation flag in the basic operations of binary floating point. Decimal floating point has a few extra cases around its special operations.

Operation	Result
$\infty - \infty$	NaN
$\pm\infty \times \pm 0.0$	NaN
$\pm\infty / \pm\infty$	NaN
$\pm 0 / \pm 0$	NaN
FMA containing an invalid multiply or add	NaN
remainder$(x, \pm 0)$ for any non-NaN x	NaN
remainder(∞, x) for any non-NaN x	NaN
$\sqrt{x}$ with $x < 0$	NaN
integer$(\pm\infty)$	No result
integer(NaN)	No result
integer(x) with $x <$ INT_MIN if rounding down	No result
integer(x) with $x >$ INT_MAX if rounding up	No result
Signaling comparisons with NaN operands	No result
intLogB$(\pm\infty)$	No result
intLogB(± 0)	No result
intLogB(NaN)	No result
Any operation with sNaN	NaN/No result

by the division by zero. Division by zero is not a catastrophe, but it's a case where floating point does something weird.

One other required floating-point operation can produce a division by zero exception, and it is the binary logarithm operation. The logarithm of 0 is well-defined as $-\infty$ (a mathematician would say that this happens in the limit), and this is another example of an operation on a finite input with an exact infinite result. These two cases with signs spelled out are shown in Table 3.2.

TABLE 3.2
Enumerated list of cases that produce a divide by zero exception in the basic operations of floating point.

Operation			Result
$x/0$	with	$x > 0$	∞
$x/0$	with	$x < 0$	$-\infty$
$x/(-0)$	with	$x > 0$	$-\infty$
$x/(-0)$	with	$x < 0$	∞
logB(± 0)			$-\infty$

Invalid operation and division by zero exceptions operate the same way for every floating point number format, and are unrelated to dynamic range or numerical precision. Invalid operations either produce a NaN or fail, and operations that raise a division by zero exception produce an exact infinity from finite inputs.

Other operations in math libraries can also produce invalid operation and division by zero exceptions, but they follow the same principles as the required operations. A case where non-NaN inputs produces a NaN output is an invalid operation, and a case where normal inputs produce an infinity will be signaled as a division by zero.

3.5   Overflow (OF) and Underflow (UF)

Overflow and underflow exceptions occur when operations exhaust one end of the dynamic range of the floating point format. Overflow (OF) occurs when the exact result of a calculation is past the maximum or minimum representable value in the given format. Underflow (UF) occurs when the result of a calculation is smaller than the smallest normal number but not exactly zero. Graphically, overflow and underflow occur when results land in the regions indicated in Figure 3.2.

Overflow is the simpler case. If the rounded result of an operation has a larger magnitude than can be represented in the result format, we get an overflow. Overflow doesn't necessarily need to come from arithmetic operations. A common place to find overflows is in conversion operations between wider and narrower floating point formats. If the number being converted is beyond the dynamic range of the destination format, you get an overflow. The behavior of overflow depends on the rounding mode chosen. The behavior is summarized in Table 3.3. Calculations involving infinities are not overflows, and frequently don't raise exceptions, while calculations of normal numbers that produce an infinity are always either an overflow or a division by zero. The difference between overflow and division by zero is that a division by zero exception signals a result that is exactly infinity, while an overflow signals a

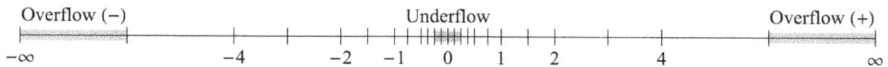

FIGURE 3.2
Regions where unrounded operation results cause underflow and overflow exceptions on an example floating point system with 1-bit mantissa and exponent range $[-2, 2]$.

TABLE 3.3
Behavior of floating-point operations under overflow, depending on the chosen rounding mode. HUGE represents the largest non-infinite floating-point number.

	Result	
Rounding Mode	**Approaching** $-\infty$	**Approaching** ∞
Nearest, Ties to Even (Default)	$-\infty$	∞
Toward Zero	$-$HUGE	HUGE
Toward Negative	$-\infty$	HUGE
Toward Positive	$-$HUGE	∞

result that could be rounded to infinity. An overflow is always also an inexact calculation, almost by definition.

It is possible to avoid creating any infinity on overflow by using the round-toward-zero rounding mode. In this rounding mode, you sacrifice accuracy since inexact calculations can be off by one ULP rather than half a ULP, but the only operations on normal numbers that create an infinity result are ones that trigger a divide by zero exception.

Technically, a floating-point underflow exception can occur in two circumstances:

1. When an inexact calculation produces a result before rounding whose magnitude is smaller than the smallest normal number.

2. When an exact calculation produces a nonzero subnormal result.

In the first case, the inexact exception is also triggered, and you get an underflow when the rounded result is zero. You also get an underflow exception when the result is rounded to a normal number, since the underflow flag depends on the pre-rounding result. This is the normal idea of what "underflow" means. The second case creates a few oddities: An operation whose exact result is a subnormal number is considered an underflow, including the exact sum of two subnormal numbers (if the result is also a subnormal number). Thankfully, the default behavior of underflow exception reporting, which is followed by most CPUs, is to only signal an underflow in the first case, when you have an inexact operation that actually *loses precision* due to the arithmetic underflow.

The definition of overflow and underflow depends only on the result of the calculation, not on the inputs, but they mostly follow the intuitive definitions of the words "overflow" and "underflow". These two exceptions are also companions to the inexact exception (excluding the non-default behavior of the underflow exception), and will generally only be signaled alongside it. They clarify two special modes in which a calculation can be inexact. For brevity, we will be omitting mention of the IX exception when an overflow or underflow occurs in an example in this book.

3.6   Exceptions and Rounding in Processors

CPUs usually have a floating point status register that indicates the exceptions quietly and a control register that allows you to set floating-point exceptions as CPU traps, interrupting computation and passing the error to a handler. There is also usually a field in the control register to set the rounding mode used. This level of flexibility allows you to convert certain types of exceptions into program-level errors and ignore other kinds.

Generally, processors will give you a lot of control over rounding modes, but limit the control you have over exception signaling. For example, they will usually not allow you to report an underflow on an exact result. However, most will let you set which kinds of errors are reported in flags and which kinds of errors cause a CPU interrupt, and some languages and programming environments take advantage of this differently than others. Almost all CPUs give you all of the rounding modes as options. It's a small amount of silicon, but has great benefits for the flexibility of programs.

Single-instruction multiple-data (SIMD) pipelines in CPUs, such as Intel's AVX, will often aggregate the signals coming from all lanes of the computation, so you will not necessarily get information about which calculation caused an error. It generally works out well to look for the NaN when you get one of these flags (or the infinity), but that does not work for some other exceptions, like the inexact exception. CPUs of the x86 family set rounding modes, control CPU traps, and report error in the MXCSR control and status register. When using an x86 CPU for floating point, if you care a lot about the exceptions in the calculation, you have to do one calculation at a time instead of using a vector [3]. ARM works much the same way, but splits the control and status registers into the FPCR for control and FPSR for status [7].

GPUs generally do not signal these exceptions in any way, and have no mechanism for converting them into traps. Since GPUs run a lot of threads at once and rely on having very little flow control for each thread, checking for and trapping on floating-point exceptions loses a lot of the benefit of using a GPU. GPUs are still compliant with the floating point standard, and will operate correctly for overflows and invalid operations, but they do not have any out-of-band reporting mechanisms for exceptions. Specialized types of arithmetic that rely on seeing exceptions (e.g., arithmetic that needs a non-default rounding mode, and needs an inexact signal) will not be able to be translated to a GPU. On the positive side, GPUs will often allow you to control rounding modes as part of the operation instruction, so you have very fine-grained control of rounding. Systems like interval arithmetic, which have half of your instructions rounding to negative and half rounding to positive, can take advantage of this capability to get better performance than you might expect [12].

When you are running a numerical algorithm that needs both accuracy and speed, it is common to do development of the algorithm serialized on a CPU and then parallelize to a SIMD or GPU implementation once you have worked out that you can safely ignore any exceptions that might be raised. Reporting errors and exceptions from parallel computation systems is much harder than reporting them from systems that compute serially, and the floating-point exceptions are no exception to that rule.

Check Your Understanding

Problem 3.1. Catch them all: Using SIMD intrinsics or a SIMD library, perform a single vectorized floating-point operation on a handcrafted input (the operation and the input are up to you) that triggers all the floating-point exceptions simultaneously.

Problem 3.2. Write a program that converts integers to single-precision floating point numbers, rounding to odd mantissas for inexact conversions. Find a large number where the input, the default-rounded version, and the rounded-to-odd version are all different.

Problem 3.3. Only using integer arithmetic, construct a program that performs integer division with round-to-nearest-ties-to-even rounding.

Generate 10 000 pairs of integers with the first number between $-10\,000$ and 10 000, and the second number between -100 and 100. Compare the accuracy of your round-to-nearest code against default integer division and a double-precision quotient of the numbers. Compute a sum-of-quotients for the whole set of numbers in each mode (integer, rounded integer, and double precision) and compare the accuracy of the results.

Problem 3.4. Write a function that performs single-precision floating point sums using stochastic rounding. Stochastic rounding involves rounding inexact operations up or down with probability related to the magnitude of the tail of the number (e.g., 1.9 would round to 2 with probability 0.9 and round to 1 with probability 0.1), and must be implemented manually using a wider format to get the exact result prior to rounding.

Add up 100 000 random numbers drawn uniformly from the range $[0, 1]$ in single precision with default rounding, single precision with stochastic rounding, and double precision. Is the result with default rounding or the result with stochastic rounding closer to the double precision result?

4

The Basic Arithmetic Operations

Floating-point systems have six basic arithmetic operations: The four operations we all know and love, plus square root and an operation that combines multiplication and addition, called a fused-multiply-add (FMA) operation. All of these operations can produce an inexact result, and rely on rounding to get back to an appropriate bit width.

I think of floating point as a format built around multiplication: The simplest operations for floating point are multiplication and division. For this reason, this chapter begins with a discussion of multiplication rather than the traditional approach of starting with addition.

Floating point operations also do not have the same algebraic properties that you expect from operations like addition and multiplication. For this reason, compilers generally make the (dangerous) assumption that you know what you are doing. This also means that you have a lot of freedom to choose how accurate or fast you want your math to be.

4.1 Multiplication

Multiplication and division are the simplest and easiest floating point operations, with the fewest pitfalls.

Both multiplication and division can be computed by separating the floating point number and computing the result on each part. Remember the formula for the value of a floating point number:

$$F = (-1)^s 2^e (1.m)$$

Since all three of these terms are multiplied together, when multiplying two floating point numbers, we can separate the multiplication:

$$F_1 \times F_2 = [(-1)^{s_1} 2^{e_1} (1.m_1)] \times [(-1)^{s_2} 2^{e_2} (1.m_2)]$$
$$= [(-1)^{s_1} \times (-1)^{s_2}][2^{e_1} \times 2^{e_2}][(1.m_1) \times (1.m_2)]$$

Simplifying by the distributive property of exponentiation over multiplication:

$$F_1 \times F_2 = (-1)^{s_1+s_2} 2^{e_1+e_2} [(1.m_1) \times (1.m_2)]$$

DOI: 10.1201/9781003565543-4

This gives us a new floating point number. The sign calculation cannot affect any other part of the number, since no other part of the number has -1 as a factor, and so the new sign bit is the XOR of the signs of the inputs (addition modulo 2 is equivalent to XOR). Similarly, the exponent calculation stays in its lane computing a new power of two by addition. However, the mantissa can leak a bit: Since each of the numbers being multiplied to create the new mantissa is between 1 and 2, the result of that multiplication may be between 1 and 4. This can cause a leak of bits into the exponent field, but that is the only inter-field dependency.

The multiplication of the mantissas may need the result to be rounded. Consider the single-precision floating-point number: $(1 + 2^{-15})_F$, also represented exactly as $0F1.000000000000001_b$. The exact square of this number is:

$$(1 + 2^{-15}) \times (1 + 2^{-15}) = 1 + 2^{-14} + 2^{-30}$$

However, the 2^{-30} portion of this product represents a problem: We only have 23 mantissa bits, so we cannot put a bit in the 30th place. This means that we have an inexact result and we need to round. Rounding will take place using whatever rounding mode has been decided before. Using the default round-to-nearest mode, we get the result:

$$(1 + 2^{-15})_F \times (1 + 2^{-15})_F \stackrel{IX}{\Rightarrow} (1 + 2^{-14})_F$$

Or in other words:

$$1.000000000000001_b F0 \times 1.000000000000001_b F0 \stackrel{IX}{\Rightarrow} 1.00000000000001_b F0$$

As expected, this operation will trigger an inexact exception. If we did this example in double precision, there are enough bits for an exact result:

$$1.000000000000001_b D0 \times 1.000000000000001_b D0 \Rightarrow$$
$$1.000000000000001000000000000000001_b D0$$

In the general case, multiplication of two mantissas produces a product that has twice the precision of the original multiplication, so rounding is often necessary.

Zero (along with subnormal numbers) is often a special case for floating point multiplier hardware. In this case, the multiplier simply has to detect a zero to produce a zero result. The same goes for infinity. However, infinity times zero gives a NaN result. The special cases can be found in Table 4.1. Note that in all of the special cases, the sign bit is still dutifully calculated as the XOR of the sign bits of the operands.

Due to the addition of the exponents, multiplication can easily overflow or underflow. Multiplying lists of numbers easily result in this when you reach a block of very large or very small numbers, even when the final product of the list is representable in floating point. This is much harder to do in

TABLE 4.1
Special cases of floating-point multiplication, where x is a positive normal floating point number.

Operation			Result	Exceptions
$\pm x$	$\times$	0.0	± 0.0	
$\pm x$	$\times$	-0.0	∓ 0.0	
0.0	$\times$	0.0	0.0	
0.0	$\times$	-0.0	-0.0	
-0.0	$\times$	-0.0	0.0	
$\pm x$	$\times$	∞	$\pm\infty$	
$\pm x$	$\times$	$-\infty$	$\mp\infty$	
∞	$\times$	∞	∞	
∞	$\times$	$-\infty$	$-\infty$	
$-\infty$	$\times$	$-\infty$	∞	
$\pm\infty$	$\times$	± 0.0	NaN	IO
NaN	$\times$	Any	NaN	IO for sNaN

floating point than it is with integers, though, since floating point has such a wide dynamic range that most products of "usual" numbers won't go beyond the range of a `double`. Multiplication is also numerically well-behaved: Aside from the normal rounding errors, you will almost never get a result that isn't close to the intuitive expectation since the multiplication of the significands maintains precision.

Multiplying floating point numbers is usually as fast as integer multiplication, and is usually the fastest floating point arithmetic operation available. For this reason, floating point algorithms designed for speed will often prefer multiplication over addition.

4.2 Division

In the normal cases, division is somewhat similar to multiplication, with the addition of a reciprocal calculation:

$$\frac{F_1}{F_2} = [(-1)^{s_1} 2^{e_1}(1.m_1)] \times \left[(-1)^{-s_2} 2^{-e_2}\frac{1}{(1.m_2)}\right] = (-1)^{s_1-s_2} 2^{e_1-e_2}\left[\frac{1.m_1}{1.m_2}\right]$$

Like multiplication, the only inter-field dependency is that the mantissa calculation can adjust the exponent calculation by one, but this time it can go down.

Two methods are generally used for computing reciprocals, functional iteration and digit recurrence. Digit recurrence division uses a method similar to the paper method of long division. Bits of the significand are grouped into digits, and the quotient is computed one digit at a time, starting with the most significant bits and moving down the number. This algorithm uses a small lookup table based on the divisor and the top digit of the current remainder to determine how many times the divisor divides the top digit of the remainder, then updates the remainder. The first Pentium processors from Intel used digit-recurrence division with a digit size of two bits. The lookup table had 5 out of 1 066 entries populated with the wrong result, resulting in errors in division that accumulated over the rest of the division calculation. Modern processors will often use 3–6 bits per digit, with a correspondingly larger lookup table [13].

Functional iteration methods start with an initial approximation, usually taken from a lookup table, and use an iterative algorithm such as Newton-Raphson to refine the precision of the result. These algorithms have quadratic convergence, doubling their precision in bits for each refinement step, but the refinement step is slower and more energy-intensive than the digit-recurrence method. For this reason, digit-recurrence division is more common to see in CPUs and GPUs, but functional iteration methods are often faster [14].

As an example of how this computation is done, the most common functional iteration method for division is Goldschmidt's method, which uses the following idea:

$$Q = \frac{N}{D} = \frac{N}{D} \frac{F_0}{F_0} \frac{F_1}{F_1} \frac{F_2}{F_2} ...$$

We can set the F's so that the sequence $DF_0F_1F_2...$ converges to 1, causing the sequence $NF_0F_1F_2...$ to naturally converge to Q. When $0 < D < 1$, there is a simple formula to find the next F based on a current value of D:

$$F_{n+1} = 2 - D_n$$

For example, if we are computing with $D = 0.7$:

$$D = 0.7 \longrightarrow F_1 = 1.3$$

$$D_1 = F_1 D_0 = (0.7) \times (1.3) = 0.91 \longrightarrow F_2 = 1.09$$

$$D_2 = F_2 D_1 = (0.91) \times (1.09) = 0.9919 \longrightarrow F_2 = 1.0081$$

$$D_3 = F_3 D_2 = (0.9919) \times (1.0081) = 1.99993439 \longrightarrow F_3 = 1.0.00006561$$

This sequence will converge so that $D_\infty = F_\infty = 1$. Intuitively, at each round, we add $x\%$ to a number that is "discounted" by $x\%$. Combining the multiplicative markup with the multiplicative discount, we get closer and closer to 1. In general, with $D = 1 - x$:

$$D = 1 - x \longrightarrow F_1 = 1 + x$$

$$D_1 = (1 - x)(1 + x) = 1 - x^2 \longrightarrow F_2 = 1 + x^2$$

$$D_2 = (1 - x^2)(1 + x^2) = 1 - x^4 \longrightarrow F_3 = 1 + x^4$$

We can see that in this sequence, we converge to 1 by squaring x at each iteration. Since x is the error in our division, we are squaring the error at each step, which corresponds to doubling the number of correct bits in the result. This form of calculation does not even require us to compute the sequence of D's. We can simply start with $x = 1 - D$ and estimate the quotient as:

$$Q = N(1 + x)(1 + x^2)(1 + x^4)(1 + x^8)...$$

Since the floating point significand calculation naturally has $1 < D < 2$, we can simply divide by two, shifting bit places, for this calculation to get $0.5 < D < 1$, and shift back for the final result. In practice, the first F is often chosen from a lookup table which has a decent approximation of $1/D$, and the iteration is just run 1–3 times. Starting with $0 < x < 0.5$, you have to run 5 rounds for 24 bits of precision, but if you start with 12 bits of precision coming out of a lookup table, you only need to run one refinement round. As you might imagine, each step needs to be done with a few extra bits of precision to get a precisely rounded final result [15].

CPUs have logic dedicated to computing these reciprocals, and they are done relatively quickly, but division is a slower operation than multiplication or addition, so you will often see compilers swapping divisions for multiplications when they are exactly equivalent. You may find that it improves performance to do this in your own formulas. If you are in the situation where you are using one denominator function frequently, it can help to take a small precision loss by pre-computing the reciprocal of that denominator and storing it. A compiler will not do this automatically, since it causes a loss of precision.

Many CPUs will also have an instruction that quickly computes an approximate division. This is a straightforward compromise of speed and accuracy, allowing you to get somewhere between 8 and 14 bits of precision in the same amount of time that it takes to compute a multiplication. Note that this problem is actually easier than integer division: since we are dividing two numbers between 1 and 2, the result is guaranteed to be between 0.5 and 2, and the precision is lower than would be needed for 64 bit integer division as well.

Like multiplication, division is often inexact, but is well-behaved. The inexact nature of division is intuitive, thinking about integer division, but the general case also would require $2n$ bits to precisely represent the significand of a division result with n bit significands. As a sister operation to multiplication, division is also prone to overflow and underflow (albeit much less prone to overflow than integers), but is similarly well-behaved. It is hard to reach overflow with "typical" numbers, and the precision of results is perfectly maintained aside from rounding, which results in a precision loss of half a ULP with rounding to nearest or one ULP with directed rounding.

The special cases are where division stands out, outlined in Table 4.2. Division by zero does not create numerical problems, as we do have an appropriate infinity to cover those cases, but it signals an exception in case you weren't expecting to divide by zero. Some languages, however, will take this exception as an actual error: C and C++ will not, although they can have

TABLE 4.2
Special cases of floating-point division.

Operation		Result	Exceptions
±0.0 /	x	±0.0	
±0.0 /	$-x$	∓0.0	
±0.0 /	∞	±0.0	
±0.0 /	$-\infty$	∓0.0	
$\pm x$ /	∞	±0.0	
$\pm x$ /	$-\infty$	∓0.0	
$\pm\infty$ /	x	$\pm\infty$	
$\pm\infty$ /	$-x$	$\mp\infty$	
$\pm x$ /	0.0	$\pm\infty$	DZ
$\pm x$ /	-0.0	$\mp\infty$	DZ
$\pm\infty$ /	0.0	$\pm\infty$	DZ
$\pm\infty$ /	-0.0	$\mp\infty$	DZ
$\pm\infty$ /	$\pm\infty$	NaN	IO
±0.0 /	±0.0	NaN	IO
NaN /	Any	NaN	IO for sNaN
Any /	NaN	NaN	IO for sNaN

some pitfalls—division by floating point zero is undefined behavior on platforms that do not conform with IEEE 754 [16]—but Python is known for turning this acceptable case into an error. Division by an infinity generates a corresponding zero, and is a completely valid operation. Division also carefully preserves the sign that would come from the operation, so the sign bit of the resulting operation is generally the XOR of the sign bits of the input operands, even when a 0 or infinity result is generated. While multiplication had one class of invalid NaN-producing operations, division has two: ∞/∞ and $0/0$ for all signs. These are analogous to the $\infty * 0.0$ case.

Unlike integer division, the floating-point remainder operation is a separate operation from division, and we will discuss its variants later in Section 7.1. The idea of a remainder feels somewhat more synthetic when we are not constrained tightly to the integers, and depends on integer conversion. Still, floating point also has an optional remainder operation, which can be set up to work like an integer remainder (see Section 7.1).

4.3   Addition and Subtraction

Addition and subtraction are the operations that have the largest number of pitfalls in floating point, and are also probably the most commonly used operations. Addition and subtraction, when you are not careful, can swallow numbers wholesale, and are also prone to a problem known as *catastrophic cancellation*, when subtracting two numbers of similar magnitude. However, it is possible to avoid these problems by learning how floating point addition works.

The algorithm used inside floating point units for addition involves justifying the significand of the lower-magnitude number with that of the higher-magnitude number by shifting right, negating as appropriate, and then adding the significands like integers. The exponent is then renormalized based on the computed sum of the two significands, possibly going up by one, going down by the precision of the numeric format, or going to zero. The output of an addition or subtraction can have many different exponents depending on the difference of the justified significands, and it is possible to have either sign depending on the magnitudes of the operands. Given two numbers to add in scientific notation, you would likely do a similar algorithm or abandon scientific notation altogether, since the notation gets in the way on addition.

To illustrate the problems of addition and subtraction, we will start with a few examples. Our first example is relatively tame, but foreshadows what is to come:

$$253_F + 4_F \Rightarrow 257_F$$

In our scientific notation, these numbers look like:

$$1.9765625F7 + 1.0F2 \Rightarrow 1.00390625F8$$

Executing the algorithm of aligning the significands, we get the following process of addition:

$$
\begin{array}{rcl}
1.1111101 & \times & 2^7 \\
+ \quad 1.00 & \times & 2^2 \\
\hline
1.1111101 & \times & 2^7 \\
+ \quad 0.0000100 & \times & 2^7 \\
\hline
10.0000001 & \times & 2^7 \\
\hline
1.00000001 & \times & 2^8
\end{array}
$$

We first normalize to the magnitude of the highest-magnitude operand, and then we add as normal. A final step is to renormalize the final number.

We can see two immediate differences with multiplication here. First, the computation cannot be done with the fields separated. The significands are where the entirety of the action is. However, they cannot just be added naively, and they have to be justified based on the exponents of their numbers for the addition. This can create a slightly uncomfortable circumstance, illustrated by our next example:

$$16777216_F + 1_F \Rightarrow ?$$

Exposing the values of the exponent and significand,

$$1.0F24 + 1.0F0 \Rightarrow ?$$

Justifying and computing the sum in binary, we get the following result:

$$
\begin{array}{rl}
& 1.00000000000000000000000 \times 2^{24} \\
+ & 0.00000000000000000000001 \times 2^{24} \\
\hline
& 1.00000000000000000000001 \times 2^{24}
\end{array}
$$

There seems to be no problem here until we count the number of zeros. There is one too many, so we need to round. We are rounding to nearest with ties going to even numbers, so the complete calculation is:

$$
\begin{array}{rl}
& 1.00000000000000000000000\ 0 \times 2^{24} \\
+ & 0.00000000000000000000000\ 1 \times 2^{24} \\
\hline
\textbf{Round}\quad & 1.00000000000000000000000\ 1 \times 2^{24} \\
\hline
& 1.00000000000000000000000 \times 2^{24} \quad \texttt{IX}
\end{array}
$$

The addition has completed without changing the first operand. In other words:

$$
1.0F24 + 1.0F0 \stackrel{\texttt{IX}}{\Rightarrow} 1.0F24
$$

In this case, adding two integers results in an unchanged value and only an often-ignored inexact exception. The default rounding mode is to round to nearest with ties to even, and the nearest even value to a mantissa of $\left[0 + \frac{1}{2}\text{ULP}\right]$ is 0. The value of 24F1.0 completely dominates this calculation, and when rounding to nearest, we round back to that operand.

This is why storing counters in floating point numbers is a bad idea. The significand of floating point numbers is substantially smaller than the width of an integer, but integer counters also can roll over once they reach 2^{32}. Floating point counters will make their way to 2^{24} and stop (assuming default rounding). Double-precision counters have a similar ceiling at 2^{53}. If you must make a counter using floating point numbers, force it to roll over at 2^{24} or 2^{53} to avoid rounding-related issues, and do modular arithmetic if you want to compare counter values.

Our final example is an example of a subtraction:

$$
16777216_F - 16777212_F \Rightarrow 4_F
$$

Or alternatively:

$$
1.0F24 - 1.9999995231628418F23 \Rightarrow 1.0F2
$$

In this form, it is clear what is happening here. We are subtracting a number that is almost equal to 2^{24} from 2^{24}, which naturally gives us a small number. The exponent of 24 that we had in both operands goes out the window, since almost every bit of the subtraction of the significands cancels to 0. In binary, that subtraction goes like this:

$$
\begin{array}{rl}
& 1.00000000000000000000000 \times 2^{24} \\
- & 0.11111111111111111111110 \times 2^{24} \\
\hline
& 0.00000000000000000000010 \times 2^{24}
\end{array}
$$

Everything about taking the exponent of the larger-magnitude number has gone. We have two numbers close to 2^{24}, and the difference between them is on the order of 2^2. We have essentially reduced from two numbers with 24 significant bits (the binary analogue of significant digits) down to one number with two significant bits. This is even worse than you might think. The real numbers corresponding to those floating point numbers could have been 16 777 217 and 16 777 211.5. We have computed a difference of 4 when the true difference is 5.5, a 37.5% error in the difference.

Alternatively, 1 could have been added to the 16 777 216 operand 10 times prior to the subtraction. The nearly insignificant error of dropping those 10 additions when the number was 2^{24} becomes a major accuracy problem when the number is close to 1. This is catastrophic cancellation, when cancellation during an addition or subtraction causes you to lose a large amount precision. 16 777 217 and 16 777 211.5 are within 0.0005% of their respective floating-point representations, but their difference is 37.5% off when calculated using floats.

It is very easy in your mathematical code to run into issues with both catastrophic cancellation and operand swallowing. The most basic example is when you are keeping a running sum of samples, and then comparing that running sum to an expected outcome. Floating point addition is often more precise than integer addition when the numbers are of a similar magnitude, but it can be disastrously imprecise when the two operands have a large difference in magnitude, as could happen during a running sum. Most numerical code uses double-precision floating point, which has enough precision to absorb most of the problems we have mentioned in this chapter without issue. Loss of precision from addition and subtraction, however, is still a risk.

Floating point addition and subtraction have relatively few special cases, enumerated in Table 4.3. Most of these cases are what you would expect. With a 0 operand, you pass the other operand. With an infinity, the infinity takes over, and the only invalid operation is the sum of infinities of opposite signs. There is a subtlety with the zeros, which is that the sum of a positive and negative zero is a positive zero. This is a way to eliminate negative zero in any calculation: You can simply add 0.0 to any calculation, and every number other than −0.0 will be unaffected, but a negative zero will become positive. In fact, the only case where a negative zero is produced is when two negative zeros are added. The sum of a number and its negative is always positive zero. Finally, the only invalid operation is the sum of infinities of opposite signs.

Despite being heavily optimized, floating point addition is often slower than floating point multiplication in microprocessors. This is natural due to the larger size of circuits required and the increased number of special cases. It is still a relatively fast operation, but not at the scale of integer addition, which is trivial by comparison. Floating point is a format made for multiplication.

TABLE 4.3
Special cases of floating-point addition. When rounding toward negative, the starred results are instead -0.0.

Operation			Result	Exceptions
$\pm x$	$+$	± 0.0	$\pm x$	
x	$+$	$-x$	0.0^*	
0.0	$+$	0.0	0.0	
0.0	$+$	-0.0	0.0^*	
-0.0	$+$	-0.0	-0.0	
$\pm x$	$+$	∞	∞	
$\pm x$	$+$	$-\infty$	$-\infty$	
± 0.0	$+$	∞	∞	
± 0.0	$+$	$-\infty$	$-\infty$	
∞	$+$	∞	∞	
$-\infty$	$+$	$-\infty$	$-\infty$	
∞	$+$	$-\infty$	NaN	IO
NaN	$+$	Any	NaN	IO for sNaN

4.4 Square Root

Many users of floating point are surprised to see that square root is given a prominent place as a required operation, when the other transcendental functions are not. Calculation of square roots is important for graphics and computations that involve geometry, like simulations, possibly more so than division. Thankfully, computing square roots in floating point is efficient and fast, on par with the speed of calculation of a division, and square roots are a well-behaved floating point operation.

Square root, like multiplication and division, can be computed on a separated form of the number. In this sense, it is easier than addition or subtraction. In practical implementations, square root calculation tends to share a lot of hardware with division. When sign, exponent and mantissa are separated, we get the following property:

$$\sqrt{(-1)^s 2^e (1.m)} = \sqrt{(-1)^s} \sqrt{2^e} \sqrt{1.m}$$

Since square roots of negative numbers are undefined in floating point, all negative square roots, with the exception of $\sqrt{-0}$, become NaN. The complete special cases are shown in Table 4.4. As expected, infinity preserves itself, and NaNs propagate.

One oddity of floating point is that unlike all the rest of the negative numbers, $\sqrt{-0} = -0$. This is for the benefit of interval arithmetic, where expressions are computed over ranges rather than single numbers. Interval

TABLE 4.4

Special cases of floating-point square root.

Operation	Result	Exceptions
$\sqrt{\pm 0.0}$	± 0.0	
$\sqrt{\infty}$	∞	
$\sqrt{-x}$	NaN	IO
$\sqrt{-\infty}$	NaN	IO
$\sqrt{\text{NaN}}$	NaN	IO for sNaN

arithmetic frequently uses the signs of zero to indicate open and closed intervals containing zero, and since most other floating-point operations will diligently preserve signs of zero, square root does as well. This is also part of why thinking about negative zero as a "negative epsilon" is not quite correct. Negative zero is zero, but we approached it from the negatives.

The positive square roots are then computed using a nice identity:

$$\sqrt{2^e} \times \sqrt{1.m} = 2^{e/2} \times \sqrt{1.m}$$

The calculation of $e/2$ can result in a non-integer, but if we pull in the least significant bit of e to the computation of the square root of the mantissa, we can make that problem go away:

$$\sqrt{2^e} \times \sqrt{1.m} = \begin{cases} 2^{e/2} & \times & \sqrt{1.m} & \text{even } e \\ 2^{(e-1)/2} & \times & \sqrt{2(1.m)} & \text{odd } e \end{cases}$$

The square root can then be approximated using a range between 1 and 4. These square roots are generally computed using one of the two iterative methods used for computing divisions, often sharing significant hardware resources with the division calculation. Digit recurrence methods use a different lookup table for square roots and a different function for functional iteration, but the overall methods are the same, either going digit by digit or refining a guess.

Square root is probably the most well-behaved floating point operation in terms of its exceptional behaviors. When $x < 1$, $\sqrt{x} > x$, meaning that the square root calculation cannot underflow. Similarly, when $x > 1$, $\sqrt{x} < x$, meaning that the square root calculation similarly cannot overflow. In many cases, such as $\sqrt{2}$, square roots will be inexact and will need to round, but you will never see an infinity from square root, and you will never see a NaN unless you pass a negative number or a NaN in. Inexact square roots are common, though, as many are irrational.

Square root is a first-class citizen in floating point arithmetic. Square root is important for graphics calculations, simulations, and some machine learning calculations, and many programs will take more square roots than divisions. It is given the same level of care by processor architects as the four basic arithmetic functions.

4.5   Fused Multiply Add (FMA)

FMA is the most recent basic operation standardized in the IEEE 754-2008 specification, but has been a useful tool of scientific computing for several decades. The idea of FMA is to compute $AB + C$ *without* doing intermediate rounding, fusing a multiplication to an addition. Essentially, the multiplication of A and B is completed with exact precision, and the result is added to C, with only a final rounding step.

Consider the example in single precision floating point where $C = -1985783424_F$, $B = 1985783424_F$, and $A = 1.00000000000000000000001_bF0$. In this case, B and C have been chosen such that B and C are large in magnitude, but able to be represented exactly. AB is very slightly larger in magnitude than C, but we have picked A such that the product AB would have an inexact result when computed alone. Breaking down the operation, the intermediate clearly cuts off several bits:

$$1.00000000000000000000001_bF0 \times 1985783424_F \stackrel{\mathrm{IX}}{\Rightarrow} 1985783680_F$$

and the sum operation gives the following result:

$$1985783680_F + (-1985783424_F) \Rightarrow 256_F$$

When computed in double precision, which has twice as many bits of precision, we can see that:

$$1.00000000000000000000001_bD0 \times 1985783424_D \Rightarrow 1985783660.723831_D$$

$$1985783660.723831_D + (-1985783424_D) \Rightarrow 236.723831_D$$

With single-precision FMA, however:

$$\mathrm{FMA}(1.00000000000000000000001_bF0, 1985783424_F, -1985783424_F)$$
$$\Rightarrow 236.723831_F$$

Since the multiplication is done at infinite precision, there is no inexactness or rounding. We have essentially prevented a catastrophic cancellation issue.

While double precision works for this example, there are analogous examples where double precision fails, and an FMA operation will give a correct result. FMA does take extra hardware and is a bit harder for embedded CPUs to compute, but it is a very good tool for numerical accuracy. By having an effectively unlimited precision for the multiplication, FMA precision issues in the addition step. We get perfect precision in situations that would otherwise have had a catastrophic cancellation, like this one, and it improves accuracy significantly when the term being added is smaller than the multiplication result.

FMA has the same sets of exceptional conditions as multiplication and adding. It can overflow, it can underflow, and it can produce inexact results. The special cases are also the same as the combination of the respective operations. If the product produces an infinity of the opposite sign as an infinity in the sum term, the FMA can produce an invalid operation exception. There is one interesting special case of an invalid operation, though. An implementation is allowed but not required to raise an invalid operation exception in the following case:

$$\mathtt{FMA}(\infty, 0, \mathtt{NaN}) \Rightarrow \mathtt{NaN}$$

Since the product of ∞ and 0 would give an invalid result exception, we are allowed to produce one, but the valid sum of this result with a NaN operand is allowed to "swallow" the exception.

Many compilers will automatically infer an FMA when the $AB+C$ pattern shows up in code. This usually helps, but it can explain why the math isn't working the same way from one build to the next. If you were expecting there to be intermediate rounding and there is no rounding, you will get a different answer than you expect. Counterintuitively, the FMA is actually doing better math, but we can sometimes see it as error because it changes the result of a calculation.

However, the FMA is a powerful addition to the family of floating-point operations. While it takes longer to perform an FMA than to perform a product or a sum, the hardware for FMA can be optimized so that the combined operation without intermediate rounding takes less time than the pair of operations with intermediate rounding. Therefore, when vendors benchmark the performance of their hardware in FLOPS (floating-point ops per second), they will frequently use back-to-back FMAs as their chosen benchmark. Each FMA counts as two "operations" in the historical context of the benchmark but uses a single fused instruction in most processors [17].

4.6   The Algebra of Floating Point

Most of the time we do math, we expect to be able to do algebra, allowing us to simplify complicated expressions into simpler ones. Mathematically, the ability to do this depends on a few properties, only some of which are provided by floating point. In practice, this means that a compiler or programming environment will faithfully replicate the order of operations that we choose when we write code.

A basic property of addition and multiplication is the commutative property:

$$A + B = B + A$$
$$A \times B = B \times A$$

Which essentially means that neither operand is special - swapping the left and right sides of the operator is possible. A second property is the associative property, the ability to reorder operators:

$$(A + B) + C = A + (B + C)$$

Reordering operations is very powerful for computers, and is generally possible with integer math. It also allows the reduction of constants, for example. The expression $2 + x + 7$ would normally be able to be evaluated as $x + 9$, but only if the commutative and associative properties hold:

$$(2 + x) + 7 \stackrel{\text{COMMUTE}}{\longrightarrow} (x + 2) + 7 \stackrel{\text{ASSOCIATE}}{\longrightarrow} x + (2 + 7) \stackrel{\text{EVALUATE}}{\longrightarrow} x + 9$$

Simplification of constants is something that compilers and humans are very comfortable doing, and is completely safe, since you are simply pre-evaluating that branch of the expression. A final property is the distributive property, which specifies the interaction between addition and multiplication. It guarantees that you can "distribute" a multiplication across the elements of an addition:

$$A(B + C) = AB + AC$$

Or more famously:

$$(A + B)(C + D) = (A + B)C + (A + B)D = AC + BC + AD + BD$$

This property allows you to factor or expand an expression that is being evaluated. Factoring reduces the number of operations that are required, while expanding allows more calculations in parallel. These algebraic properties are also part of how compilers optimize integer math. Analogous properties exist in real arithmetic for subtraction and division, treating subtraction as an addition of a negative and division as a multiplication of a reciprocal. Integer division breaks this property, but with addition, subtraction, and multiplication, you can factor and expand integer calculations and reach the same results.

If you have been following along with the ways floating point addition swallows precision, you might suspect that in floating point, the algebra you can do without affecting the outcome of your expression is much more limited. Thankfully, the commutative property still holds. Both operands are treated equally. Let's consider the associative property:

$$(A + B) + C \stackrel{?}{=} A + (B + C)$$

As you may have gathered from our discussion of addition, floating point addition is not associative. Consider the case in single-precision floating point using the default round-to-nearest rounding mode with $A = 2^{24}$, $B = 1$, and $C = 2^{-2}$. We get:

$$(A + B) + C = (1.0\text{F}24 + 1.0\text{F}0) + 1.0\text{F}{-}2$$

The operation in parentheses is inexact. In this case, it rounds toward the nearest even value, as we only have 23 bits of precision and the exact form would have a mantissa of $1 + 2^{-24}$, giving us half of a ULP to round. Since we are exactly between two mantissas, we choose the even one. The result is:

$$(A + B) + C = (1.0\text{F}24 + 1.0\text{F}0) + 1.0\text{F}{-}2 \overset{\text{IX}}{\Rightarrow} 1.0\text{F}24 + 1.0\text{F}{-}2 \overset{\text{IX}}{\Rightarrow} 1.0\text{F}24$$

When we change the order, though, we get the following calculation for the first step, with an exact result:

$$A + (B + C) = 1.0\text{F}24 + (1.0\text{F}0 + 1.0\text{F}{-}2) \Rightarrow 1.0\text{F}24 + 1.25\text{F}0$$

In this addition, we would create an exact mantissa of $1 + 2^{-24} + 2^{-26}$, leaving us with slightly more than half a ULP to round. Since we are rounding to nearest, we round up to one ULP, leaving us with the result:

$$A + (B + C) = 1.0\text{F}24 + (1.0\text{F}0 + 1.0\text{F}{-}2) \Rightarrow 1.0\text{F}24 + 1.25\text{F}0$$

$$1.0\text{F}24 + 1.25\text{F}0 \overset{\text{IX}}{\Rightarrow} 1.00000000000000000000001_b\text{F}24$$

This is not the same as the result we got doing the addition in the other order! Both of these results are equally wrong when considered in the infinite-precision real number case, because both are off by two. However, the first sequence had two inexact results, and thus two rounding steps, while the second sequence had only one inexact result. The preservation of the extra few bits meaningfully changes the result.

Although this is a contrived example, this loss of precision is a real problem that occurs when summing long vectors of floating point numbers. Large numbers can swallow smaller numbers while either losing bits or going completely unchanged. One option is to sort by magnitude before taking a sum and then add from smallest magnitude to largest magnitude, but this is not foolproof, especially when negative numbers are involved.

Similarly, due to rounding between computations, floating point multiplication is not associative either:

$$A \times (B \times C) \neq (A \times B) \times C$$

Multiplication does not have the same chance for "error explosion" as addition when done over a large list of numbers, but each operation rounds up or down as needed, dropping half a ULP of precision each time. Another multiplicative mathematical identity, the ability to distribute floating-point square roots over floating-point multiplication, also does not hold for the same reason of inexactness and rounding:

$$\sqrt{A \times B} \neq \sqrt{A} \times \sqrt{B}$$

The counterexamples to these are somewhat more subtle. In both cases, the trouble occurs any time you have a set of inexact multiplications to do, meaning numbers that are either large or otherwise need a lot of precision (specifically in binary—0.1 is a "precise" number). In the case of square roots, the expression on the right side rounds three times, while the expression on the left will only round at most twice. In the case of multiplication, you can get an instance where one ordering has two inexact operations and the other ordering is inexact and then exact. It is also far easier to run into issues with these two types of operations in the non-default rounding modes.

Finally, due to rounding and cancellation, the distributive property does not hold between multiplication and addition:

$$A \times (B + C) \neq A \times B + A \times C$$

Completing the chapter with another contrived example bordering on the absurd, we will look at a different exceptional condition that breaks these algebraic properties, and an example where cancellation helps. Consider double precision arithmetic with $A = 256_D$, $B = 1.000000000000000000000001_bD1020$, and $C = -1020D1.0_b$. With numbers close in magnitude to 2^{1020} but with alternating signs, we know we have some catastrophic cancellation coming. Working in factored form, we do see a large cancellation:

$$1.0D8 \times [(1.000000000000000000001_bD1020) + (-1.0D1020)] \Rightarrow 1.0D8 \times 1.0D1000$$

$$1.0D8 \times 1.0D1000 \Rightarrow 1.0D1008$$

However, this is a sane result. It does not raise any floating point errors, and it does produce a precise result (as long as we actually intended to subtract numbers near 2^{1020}). The calculation on the expanded form, however, takes a very different turn:

$$1.0D8 \times (1.000000000000000000001_bD1020) \stackrel{\mathtt{OF}}{\Rightarrow} \infty_D$$

We can't represent $2^{1028}(1 + \epsilon)$, since the maximum exponent of **double** is 1023. The only thing we can do is overflow to ∞. Similarly,

$$1.0D8 \times (-1.0D1020) \stackrel{\mathtt{OF}}{\Rightarrow} -\infty_D$$

The pièce de résistance is the final operation for the expanded calculation,

$$1.0D8 \times (1.000000000000000000001_bD1020) + 1.0D8 \times (-1.0D1020) \stackrel{\mathtt{OF}}{\Rightarrow} \infty_D - \infty_D$$

$$\infty_D - \infty_D \stackrel{\mathtt{IO}}{\Rightarrow} \mathtt{NaN}_D$$

There are inexact examples of the distributed property going wrong as well, but it is fitting to finish off with the most catastrophic cancellation possible, from infinities to a NaN. This final example seems contrived, but if you are computing the average of a large set of numbers, you may find an example of exactly this kind showing up naturally.

There is very little algebra that your compiler can automatically do with your floating point numbers, and most of the algebra you do will have some affect on the result. Whether you are thinking about speed or accuracy, the onus is on you, the user, to find the best formulas to use. Doing a little bit of algebra and thinking about the relative sizes of your numbers can pay dividends in terms of both accuracy and avoidance of NaNs.

Check Your Understanding

Problem 4.1. Write a program that computes floating point multiplication using only integer arithmetic, and without using built-in float types. Check your program with 1 000 random multiplications to verify correctness. Check your program's overflow and underflow behavior.

Problem 4.2. Write a function that performs a subtraction while tracking the precision loss due to cancellation. Treat the inputs to the subtraction as having error of $\pm\frac{1}{2}$ ULP of uncertainty, and return two outputs: The difference between the inputs and the uncertainty in the result. Subtract two numbers near 10 000, what is the uncertainty of the result? Subtract two numbers near 10 000 000 000 and report the uncertainty of the result. Generate numbers randomly between 1.0 and 1.01, and multiply by 2^{10}, 2^{20}, 2^{30}, and 2^{50}. What is the average uncertainty at each range? How does that uncertainty relate to the magnitude of the inputs and to the magnitude of the results?

Problem 4.3. Write a program that takes the double-precision sum of the following list:

```
[-10000000000000.0, -23597.0, -5.0, -0.00003,
  0.00000006, 0.01, 2.7, 16.0, 965.0, 86437628.0,
  2125040692162.0, 7874872892830.0]
```

What is the result you get when you take the sum in the printed order? When computed in different orders, what is the difference between the maximum and minimum results? What is the correct sum of this list?

Problem 4.4. Find a rounding mode and a value of x in double-precision floating point where $(2 + x) + 7 \neq x + 9$.

Problem 4.5. Without using overflow or underflow, find an example of a rounding mode and three floating point numbers where multiplication does not associate: $A \times (B \times C) \neq (A \times B) \times C$.

Hint: They will be easier to find if you avoid using round-to-nearest.

Problem 4.6. Write a program that computes correctly rounded single-precision floating point division using only integer arithmetic. Randomly generate 10 000 floating point numbers in the range $[-1\,000, 10\,000]$ and check your work against your computer's built-in floating-point division unit using default rounding. Generate 10 000 floating point numbers in the range $[0, 1]$ and check with rounding up. Check all of the special cases from Table 4.2.

Challenge version: Do this without using integer division.

5

Comparing Floating-Point Numbers

Comparisons of floating-point numbers does not create inaccuracy in itself, but the interface between the world of floating point and the world of other computations is where inaccuracies can solidify into odd behavior. Comparisons are the primary interaction with software control flow, and landing on one side or the other of a comparison can result in vastly different computations to occur. Comparisons can also be a place where exceptional conditions become problems in a calculation.

It is common knowledge that you should not directly compare the equality of floating point numbers, and you should always use some "epsilon" value. This is due to inexactness and rounding in the previous computation chain. To set that epsilon, we may have to look backwards at what we have done before in order to see how bad things could have gotten.

5.1   Relative Comparisons

Greater than, less than, greater-or-equal, and less-or-equal are relative comparison operations. These are the most well-behaved operations in floating point, but the floating point numbers are not all ordered with respect to each other. There are four possible relations between floating-point numbers, **less than**, **equal**, **greater than**, and **unordered**. When one or both members of a comparison are NaNs, the two numbers are unordered, but the numbers otherwise compare as you would expect.

This means that there are actually eight different ways to compare the relative magnitude of floating-point numbers, shown in Table 5.1. Unlike with integers and real numbers, when a floating-point number is not less than another floating-point number, that doesn't mean that it is necessarily greater than or equal to that number.

In code, these are all usually represented in the ways you would expect for comparing numbers, but it is a little unusual for the comparison (a <= b) to not return the same result as !(a > b). This does hold, however, if neither a nor b can be a NaN. This both means that your compiler cannot optimize the latter comparison to the former, but also that you may get some benefits from using a weirder-looking comparison when NaN may be an input.

DOI: 10.1201/9781003565543-5   54

TABLE 5.1

Relative comparison operations on floating-point numbers and the conditions under which they are true.

Comparison		True If
Greater than	$a > b$	greater
Less than	$a < b$	less
Greater or equal	$a \geq b$	greater, equal
Less or equal	$a \leq b$	less, equal
Not greater or equal	$a \ngeq b$	less, unordered
Not less or equal	$a \nleq b$	greater, unordered
Not greater than	$a \ngtr b$	less, equal, unordered
Not less than	$a \nless b$	greater, equal, unordered

Relative comparison with infinities also works as expected. Positive infinity is greater than any number other than another positive infinity, and negative infinity is less than any other number. The one unintuitive relative comparison is:

$$(-0.0_F < 0.0_F) \Rightarrow \texttt{false}$$

Negative zero is not less than positive zero. The two zeros are treated as equal, no matter how they are derived. Remember that the zeros are actually zero—not just indications of underflow—and the comparisons reinforce that. If you happen to need negative zero to be less than positive zero, the standard comparisons will not do that, and you will need to use an operator referred to as "totalOrder" that imposes a total ordering on the floating-point numbers (see Section 5.7).

It is common to account for inexactness in a calculation chain by adding an "epsilon" value to comparisons. Relative comparisons do not need these. Inexactness of operations can result in comparisons that "should" go one way going another way, but landing on one side or the other of that comparison is likely still a valid result. For example, a collision in a video game at zero velocity often has the same effective result as a near-miss with zero velocity.

As we will see later in the chapter, relative comparisons can create bugs in programs when you both combine the comparison with other numerical operations and take very different code paths on one side of the threshold compared to the other. These are places where you might want to consider how to handle your near-misses. However, the use of an epsilon value is usually not the way to do this for relative comparisons, as it simply adjusts the position of the threshold. For every epsilon, there is going to be a new class of near-misses.

Instead, the analogue to an epsilon for relative comparisons is a dead zone, and does not apply to many situations (unlike an epsilon, which nearly all equality checks can use), because a dead zone involves giving your binary comparison a third result: You will have a region which is `true`, a region which is `false`, and the dead zone, in which the result is treated as indeterminate.

This is a hybrid of two comparison operations, using a small epsilon in one or both directions:

$$\text{deadZone}(x < y) = \begin{cases} \text{true} & x < y - \epsilon_l \\ \text{indeterminate} & y - \epsilon_l \leq x \leq y + \epsilon_h \\ \text{false} & x > y + \epsilon_h \end{cases}$$

This only helps if you can handle an indeterminate result. For comparisons that are truly binary, like collisions and most thresholding operations, there is little point to using a dead zone.

5.2　Minimum and Maximum

An extension of comparison is taking the minimum and maximum of a set of numbers. It is common to use a comparison operator and a conditional statement to choose the minimum or maximum number, as below:

$$\min(x, y) \approx \begin{cases} x & y > x \\ y & \text{otherwise} \end{cases}$$

This is how many languages construct `min` and `max` when given a comparison operation but no other direction. It normally works. It works when x and y are numbers, and even works when one of x or y is an infinity. As long as there is an order between x and y, this operation works as expected. However, when one of your numbers is a NaN, the comparison operation may not return the result you expect. Recall that every comparison with a NaN returns false because NaNs are not ordered. Thus, when x or y is NaN, the formulation above will always give x. We can make an alternative formulation that will always give y if there is a NaN by flipping the comparison:

$$\min(x, y) \approx \begin{cases} x & x \leq y \\ y & \text{otherwise} \end{cases}$$

Since $x \leq y$ is always false when x or y is a NaN, this function always yields y when x is a NaN.

To cover the cases, floating point specifies operations for minimum and maximum that handle NaNs intelligently. If either operand is NaN, the `minimum` and `maximum` operations will return NaN. This is not dependent on the order of operands:

$$\text{maximum}(\text{NaN}_F, -55_F) \Rightarrow \text{NaN}_F$$

$$\text{minimum}(-55_F, \text{NaN}_F) \Rightarrow \text{NaN}_F$$

An interesting quirk of these operations is also that they treat negative zero as being less than positive zero, so:

$$\text{minimum}(-0.0_F, 0.0_F) \Rightarrow -0.0_F$$

This is the behavior of the minimum and maximum functions in the math libraries for Golang, Python, and Javascript [8].

An alternative form of min and max allow you to remove NaNs from calculations. The `minimumNumber` and `maximumNumber` operations, will return the number if one operand is NaN and the other is a number. This is the opposite of what you would see from other operators, but allows the removal of NaNs from calculation chains. If both arguments are NaNs, this operation gives a NaN result because it is unavoidable. It is also operand-order independent. For example:

$$\mathtt{maximumNumber}(\mathrm{NaN_F}, -55_F) \Rightarrow -55_F$$

$$\mathtt{maximumNumber}(-55_F, \mathrm{NaN_F}) \Rightarrow -55_F$$

$$\mathtt{maximumNumber}(\mathrm{NaN_F}, \mathrm{NaN_F}) \Rightarrow \mathrm{NaN_F}$$

The C and C++ `fmin` and `fmax` functions implement `minimumNumber` and `maximumNumber` respectively [16].

The minimum and maximum operations are currently considered optional in floating point, so their implementations by languages are not as uniform as you might expect. A comparison-based min and max have behavior that depends on operand order, while there are two options for floating-point specific minimum and maximum depending on whether you prefer to swallow or propagate NaNs.

5.3 Equality and Epsilons

Floating point numbers can be compared for equality just using an equality operator, but the usual intent of checking for equality in floating point is not to check whether the *floating point* numbers are equal, but whether to the calculation in the *real numbers* was equal. If any part of the calculation was inexact, the equality operator will not work for that purpose. We also cannot construct structures that do this perfectly while still operating in a bounded amount of compute time. Therefore, most equality checks of floating-point numbers are done with some allowed error, denoted by ϵ (epsilon).

For completeness, Table 5.2 shows the two exact comparison operators (equal and not equal). Note that NaN is not equal to everything, including another NaN:

$$(\mathrm{NaN_F} = 2_F) \Rightarrow \mathtt{false}$$

$$(\mathrm{NaN_F} \neq \mathrm{NaN_F}) \Rightarrow \mathtt{true}$$

Even if the bits of the NaN operands are exactly the same, the two operands are still unordered. Also, the two zeros are equal:

$$(-0.0_F = 0.0_F) \Rightarrow \mathtt{true}$$

Otherwise, the two numbers are equal if and only if they are exactly equal.

TABLE 5.2
Exact comparison operations on floating-point numbers and the conditions
under which they are true.

Comparison		True If
Equal to	$a = b$	equal
Not equal to	$a \neq b$	less, greater, unordered

As discussed in the beginning of the section, if we have two numbers and we
would like to check whether they are equal, we can compare the numbers with
the equality operator. If one of those numbers is the result of a calculation,
however, rounding during that calculation will cause the number to deviate
from its real-valued counterpart. For this reason, most floating point equality
checks are turned into a windowed comparison with an epsilon:

$$\text{epsEquality}(x, y) = \begin{cases} \texttt{false} & x < y - \epsilon \\ \texttt{true} & y - \epsilon \leq x \leq y + \epsilon \\ \texttt{false} & x > y + \epsilon \end{cases}$$

Unlike the dead zone calculation from Section 5.1, this comparison has a
binary outcome, so it is a drop-in replacement for an equality check. What is
actually happening in this form of comparison is that we are allowing more
than one floating-point x to be "equal" to y. In other words, we have traded out
the chance of false negative comparisons due to inexactness for false positive
comparisons. With this equality check, we can have x and y calculated exactly
to be values that are within ϵ, and we will falsely say that these two results
are equal. By increasing ϵ, we accept more false positive comparisons at the
cost of avoiding more false negative comparisons.

In many systems, especially physics simulations and, by extension, game
engines, the epsilon-equality comparison is very useful, and should be the
default equality comparison. Simulations will frequently have long chains of
inexact operations, many of which are dependent on the outcome of other
inexact operations. The false negative in a simulation is often a lot more
costly than a nearby false positive, as it may result in outcomes like objects
failing to collide into each other or threshold values failing to trigger when
they are checked by equality.

If the calculation chain *was* exact and had no chance of inexactness, there
is no reason to use an epsilon. In this case, the real-valued result of the calculation is the same as the floating-point result of the calculation. This is the case
for `for` loops in languages like JavaScript, where the default representation of
integers is in floating point. As long as the integer calculation stays inside the
region $\left[-2^{53}, 2^{53}\right]$ and does not include division, it will be exact.

5.4   Setting Epsilon

Setting a perfect value of epsilon amounts to tracking the error in your calculation that might cause deviations from the real-valued result. There are three suggested methods suggested for setting ϵ that can apply in different circumstances. The first option is to determine how big of a buffer you physically want, and use that as a magic number. This buffer can be relative to the magnitude of x or y or an absolute number. This is the easiest method, but can lead to logic bugs or glitches if you guess wrong. A second option is to backtrack the error through your calculation and then use that to figure out the maximum possible value you need epsilon to be. This process is called backwards error analysis, and we will discuss it in more depth in Section 9.5, but it will give you the exact minimum value of ϵ at the cost of being a relatively onerous process. Finally, you can do some experimentation at high precision in testing to get an idea of the range of ϵ you might need in a lower precision.

The first option is far more straightforward than the other two. It may be what you do today. Some people have a favorite magic value of epsilon, but if you can set epsilon in relation to whatever you are specifically doing, you get a small bonus in accuracy for no cost. For example, if you have a 4k screen and you treat the bottom left pixel as the point (0.0) and the top right as the point $(3\,840.0, 2\,160.0)$, you may be able to get away with $\epsilon = 0.5$ for collisions involving objects drawn on that screen, and certainly $\epsilon = 0.01$ will not cause collisions to have a visual gap even with anti-aliasing. This number may be larger than you normally use, but visually has the same impact as $\epsilon = 0.000001$. They are both well below the visual resolution of the output, and the larger epsilon actually can prevent glitches like objects that seem to stick together due to the gap being less than a pixel. Finally, if you have some idea of the general magnitude of numbers you are using, you can set epsilon to make sure that it actually has an effect. Several pieces of code use the same epsilon for $x = 10\,000\,000$ as $x = 0.01$, but in order-of-magnitude terms, these have completely different meaning, and if the difference in magnitude between x and ϵ is large enough, the epsilon will have no effect.

Backward error analysis involves tracking operations backward from your comparison and adding up the possible error you get. This is similar to tracking significant figures through a calculation, except we do it backward rather than forward. Most operations account for half a ULP of error, but there are cases, like catastrophic cancellations, that account for a lot more error. A not-too-bad proxy for a real backward error analysis is simply counting the number of operations involved in producing a number. 20 operations will produce something like 10 ULPs of error, so an ϵ of 10–50 ULPs is probably enough to cover your imprecision. We are doing numerical engineering here, not numerical analysis. A more detailed explanation of backward error analysis is in Section 9.5.

Finally, the experimentation method involves using either higher-precision arithmetic or a technique like interval arithmetic and producing chains of operations that are known to be close to your threshold. This can be done by adding random jitter to an input or using interval arithmetic (Section 9.4) for an interval of values that will end up near the threshold. By trying these operations or these intervals and watching the spread of the output, you can experimentally estimate your error compared to a real-valued calculation. This allows you to statistically figure out how big your epsilon should be even when a backwards error analysis would be too difficult.

5.5 Exceptions in Comparisons

With floating-point operations that have floating-point results, exceptions other than inexactness will result in an exceptional-looking output. They produce an infinity, a zero, or a NaN. Comparisons can only produce **true** or **false**. For this reason, the floating point comparisons tend to be more aggressive about exception signaling—when supported by the CPU and programming language—than the other operations. The default relative comparisons are signaling comparisons, and will signal an invalid operation exception for *any* NaN operand.

The floating point standard also defines quiet comparisons that behave like the other operations on NaN, only signaling on sNaN. The default equality comparisons are quiet, even though the relative comparisons are signaling. Logically, any NaN will not be equal to a number, but if your operation chain produced a NaN, the magnitude of the output of that chain of operations is indeterminate. All we know is that it's not a number.

The standard includes quiet and signaling comparisons for all of the comparison operations, but most languages and systems do not use the optional quiet relative comparisons or the optional signaling equality comparisons. The min and max functions have no notion of quietness like this, and behave exactly like any other floating-point operator. They yield exceptions exactly as normal, only signaling invalid operation for sNaN inputs.

The epsilon-equality function does not have the same exceptional behavior as a standard equality comparison, and needs to be done in a specific way if we would prefer that NaN stay not equal to the numbers. To maintain $\text{NaN} \neq x$, the suggested formulation of the epsilon-equality function is also the most intuitive way to construct it:

$$\text{epsEquality}(x, y) = (y - \epsilon \leq x) \wedge (x \leq y + \epsilon)$$

with $\wedge$ representing a logical AND. Both comparisons will evaluate to `false` if either x or y is a NaN. Since it uses two relative comparisons, it will also not have signaling behavior (if that matters in your application) unless you can find a way to specialize the comparisons to be the signaling versions of the comparison.

5.6 Regime Comparisons

There are several operations that are comparisons to check the regime of a floating-point number. Most math libraries define functions to check the following about a floating-point number:

- `isNormal` checks whether a floating-point number is normal.

- `isFinite` checks whether a floating-point number is finite.

- `isInf` checks whether a floating-point number is infinite.

- `isNaN` checks for any NaN.

- `isSignaling` checks for any signaling NaN.

Each of these comparisons is a quick way to look for numbers that may otherwise be hard to differentiate. The comparison for determining a NaN from an infinity is a slightly messy thing to do, so it is better to push that down to your math library. Most of these checks are actually done by extracting and comparing individual bit fields from the floating-point number: For example, `isNormal` looks for an exponent field that is not all zeros or all ones. That means that these checks do not rely on comparison operators, which could otherwise be a messy way to perform a similar function.

5.7 Total Ordered Comparison

The floating-point numbers have several numbers that are not ordered with respect to each other. The NaNs compare as unordered and the zeros compare as equal when you use a standard comparison operator. There is a relatively esoteric comparison operator that can distinguish the ordering of these equal numbers, the **totalOrder** operator. This operator is a comparison on the floating-point numbers as though they were totally ordered. This is an operator that is rarely present in math libraries as it is mostly for decimal comparisons (the decimal floating-point numbers have many representations for the same value, while binary floating point only has the two zeros), but is possible to emulate if you need it in binary floating point.

The binary version of totalOrder operates in the following way on two non-NaN numbers, as a less-or-equal check with the numbers in a strict ordering:

$$\text{totalOrder}(x, y) = \begin{cases} \text{true} & x < y \\ \text{false} & x > y \\ \text{true} & (x = -0) \land (y = +0) \\ \text{false} & (x = +0) \land (y = -0) \\ \text{true} & x = y \end{cases}$$

Decimal floating point also has a few additional cases, since one number can be represented in multiple ways. In those cases, the total ordering of the numbers falls back on the exponent, where numbers with a smaller exponent are considered to be closer to zero.

Unlike most operators, totalOrder cares about the sign bit of NaN operands. Negative NaNs are considered less than all numbers (including negative infinity), while positive NaNs occupy the other end of the spectrum. When two NaNs are compared, negative is less than positive and signaling is considered to have a smaller magnitude than quiet, but ordering is otherwise up to the implementation.

5.8   Interactions between Comparisons

Simultaneous comparison operations can have some interesting results when one of the comparisons is a relative comparison and the other one is an equality comparison. The behavior of the relative comparison is best when it is done without an epsilon, while the equality will often need an epsilon value for best use. Remember that this bad interaction is fundamental to the idea of an epsilon-based comparison: We want these false positives in the equality check to avoid the false negatives.

This returns us to the Minecraft bug from the preface. In Minecraft, when you fall off a high cliff while your character is in a boat, you will not take fall damage unless you fall exactly 12, 13, 49, 51, 111, 114, 198, 202, 310, or 315 blocks. From those heights, you will take fall damage. At the time of writing, the game developers have known about this bug for 7 years and it still has not been fully patched [18].

The Minecraft simulation engine uses two comparisons: a collision check with the ground using a relative comparison and a check whether your character is on the ground, which uses an equality check with an epsilon. Normally, when a player hits the ground, the collision happens in the same frame at which the player's "on ground" status flips from false to true. If you are in a boat, and you are not on the ground, and you take fall damage, the fall damage will be negated by the boat.

Minecraft works in units of "blocks", and the acceleration of gravity in Minecraft is nominally 0.04 blocks per tick squared (a tick represents $\frac{1}{20}$ seconds), but when converted to floating point, 0.04 cannot be represented exactly. The actual acceleration of gravity in Minecraft is slightly less than 0.04. The error is about $\frac{1}{3}$ ULP. For one reason or another, this constant appears to be in single-precision floating point while some of the math around these calculations is in double precision.

The height of the ground is in blocks, so it is always a relatively small integer, and the frame number is also an integer. This means that if you are falling for n ticks, the speed you will have at tick n is:

$$v(n) = 0.04_\mathrm{F} n = 0.0399999991_\mathrm{F} n$$

Compared to nominal, our velocity accumulates a very small error from nominal over time. The actual fall distance is as a function of tick index is:

$$d(n) = \sum_{m=0}^{n} v(m) = 0.04_\mathrm{F} \frac{n(n+1)}{2}$$

Since we are picking up about $\frac{1}{3}$ of a ULP of error compared to the nominal velocity, our fall distance is accumulating error. In the beginning of the fall, since the velocity is smaller than position, this error is actually swallowed in the position update, but as velocity deviates more and more from the nominal velocity that would be given with 0.04 instead of 0.0399999991, the error leaks into the distance calculation as the game integrates velocity to get position (see Chapter 12).

The list from the first paragraph gives the times when the nominal fall distance is exactly an integer. If you fall for 24 ticks, just over a second, the nominal fall distance is exactly 12 blocks, but the actual fall distance is just short of 12 blocks. This causes the "on ground" comparison to trigger *the frame before* the collision-with-ground comparison, which in turn causes the behavior of the fall damage calculator to change in the following simulation tick. This is due to a false positive from the value of epsilon used for the "on ground" check. With math done entirely in double precision, the resulting value will be more than 2^{20} ULPs off, but many game developers both work in single precision, where this distance is below one ULP!

Most simple fixes will not eliminate this bug. The code can likely have a significantly reduced epsilon, but that will not eliminate the bug, but just make it harder to trigger. Even if the Minecraft developers multiplied their units by 100 so that the acceleration of gravity were exact, there would still be certain fall heights where this behavior would persist.

The core of the bug here is the use of a relative and absolute comparison together with necessarily different thresholds. Ultimately, the only way to fix this bug is to completely eliminate the interaction between the relative and absolute comparison. this is often best done by turning one comparison into a state machine based on the other—for example, having the "on ground" status toggle true on collisions with the ground and false when the player gets ϵ away from the ground. The Minecraft developers in late 2024 pushed a patch that appears to reduce the acceleration of gravity on boats after they have fallen a certain distance. This is one way to make this bug significantly more difficult to trigger.

This situation is one in which a dead zone in the relative comparison can become useful. A dead zone comparison could allow the game engine to specifically perform actions aligned with the change of the "on ground" status in the dead zone while reserving actual collision logic for the cases beyond the dead zone. Alternatively, the dead zone could defer the switch of "on ground" status to the next tick. Ultimately, these sorts of fixes are always more complicated than they seem. However, this is a good lesson in the dangers of using absolute and relative comparisons together. If you are not certain of the exactness of your calculation, you will need to make sure that you either keep thresholds the same between both comparisons or use algorithmic tricks to avoid bugs.

Check your Understanding

Problem 5.1. Taking a double-precision floating point calculation of the function:

$$f(x) = \frac{3(x+1)}{x} - \frac{5-x}{100x} + \frac{20-x}{x-3} + \frac{10+x}{5+x}$$

We would like to alert the user whenever the real-valued result of this comparison would be outside of the range $(-10, 10]$ (including undefined values). Construct the fastest possible comparison on the output of $f(x)$ that produces this alert. Test near $x = 0$, $x = 3$, and $x = -5$.

Problem 5.2. Using the bitwise representation of floating-point numbers and the floating-point comparison operators, write an implementation of the `totalOrder` function described in Section 5.7.

Problem 5.3. Assuming all calculations are done in double precision, determine the maximum epsilon that would fix the Minecraft bug for all of the fall distances mentioned in Section 5.8. Find a fall distance that would still trigger the bug.

6

Conversion to and from Floating Point

Having a beautiful numeric format means very little without the ability to move numbers into and out of it. There are two main types of conversions that matter for floating point: Conversion with integers, and conversion with strings. One is easy, and the other is more nuanced. Conversion to integers uses many of the same mechanisms we have discussed before, specifically rounding. Due to the common interplay between integers and floating-point numbers on computers, these conversions are usually done quickly in hardware, taking one instruction that is usually faster than division.

String conversion, however, often involves writing and reading a form that is shortened from the precise value of a floating-point number. Computers will produce `"0.1"` when asked to convert the floating-point number closest to 0.1 to a string, while the number 0.1 cannot be represented in floating point. This conversion is one of the more misunderstood aspects of floating point, since it leads to results that are hard to explain without realizing that string conversion is usually inexact.

6.1 Rounding to Integers

While not conversions *per se*, rounding operations behave like conversions, and are the first step in the conversions to an integer. Rounding to integers can be done in any of five rounding modes, which can be specified for each operation separate from the rounding used when floating-point operations are inexact. The available rounding modes are the four floating-point rounding modes described in Section 3.1, and rounding to nearest with ties away from zero. Rounding with ties away from zero still has the numerical problems demonstrated in Section 3.1, but we expect rounding to integers to be infrequent and intentional.

The rounding functions or the directed rounding modes are often referred to by the common names of their rounding modes:

- Rounding down to an integer is the `floor` function.

DOI: 10.1201/9781003565543-6

- Rounding up to an integer is the `ceiling` (`ceil`) function.

- Rounding toward zero to an integer is the `truncate` (`trunc`) function, representing truncating the part of the number after the decimal place.

Additionally, rounding to nearest with ties away from zero (schoolbook rounding) is usually the default **round** function in math libraries. Some math libraries, including JavaScript's, do not have a function that rounds to nearest integer with ties to even despite that this is the default rounding mode for other floating point operations [8]. All of these operations are simple enough that they are done in one hardware instruction on most processors.

The rounding operations are all well-behaved. Zeros and infinities pass through these operations untouched, and NaN operands produce a NaN result, signaling if appropriate. The one interesting behavior of rounding is that every operation that doesn't have an integer input signals an inexact exception, despite the result of the rounding operation being exact. We lost precision, so we indicate inexactness.

6.2 Narrowing and Widening Formats

Conversion between floating-point formats is relatively straightforward, but it is still a conversion. Moving to a wider format is simple, and there is no possibility of exceptional cases. The exponent gets widened and re-biased, and the mantissa gets widened and trailing zeros are added. The dynamic range of wider formats is strictly larger, so there is no chance of overflow or underflow, and every number that can be represented in a narrow format can be represented in a wide format. The only complication is the subnormal numbers, which need to be turned into normal numbers when a wider format is used, so their exponent has to be computed. Still, this operation is completely numerically safe and mostly trivial.

Narrowing floating-point numbers has a few more pitfalls. In the usual case, the exponent is re-biased and narrowed and the mantissa is rounded appropriately for the current rounding mode. If the new exponent is too large or too small for the new format, then we have an overflow or and underflow and the respective exceptional behavior occurs. If the mantissa had trailing bits that got rounded away, the conversion of sizes is inexact. Moving from a wider format to a narrow format can lose precision, and the overflows and underflows can cause problems if you're not careful.

The European Space Agency learned their lesson about narrowing conversions the hard way when the Ariane 5 rocket crashed due to an overflow. The conversion in question was from a 64-bit floating-point number to a 16-bit integer, but the same overflow and underflow problems existed as would with a narrow floating-point format. The 64-bit number exceeded 32 767, and the computer performing the calculation within the inertial measurement unit caught the exception, shutting itself down. Without this system, Ariane 5 crashed into the ground [19]. While this was a conversion to an integer format and not a conversion between floating-point formats, the same lesson applies to narrowing conversions within the floating-point formats: any narrowing of the dynamic range of a format risks overflow and underflow.

Like rounding, narrowing and widening numbers is an operation that is usually done in hardware. There are few complications to these operations, and they rely entirely on building blocks that are parts of many other operations.

6.3   Integer to Float Conversion

Converting integers to floating-point numbers is the most straightforward conversion operation, and on numbers that are a "typical" magnitude, is usually exact. Integer to float conversion is also very fast, and often done with single instructions in hardware.

Bitwise, the conversion of an integer to a floating-point number involves the following steps:

1. Set sign bit based on the sign bit of the integer.

2. Treating the integer as positive, find the leading one bit of the integer, giving us the order of magnitude of the integer.

3. Remove that bit, since it is implied, and shift appropriately to make the mantissa. When the order of magnitude is less than p, this is a left shift, and when it is greater than p, this is a right shift.

4. If we did a right shift that lost bits, round the mantissa appropriately for the rounding mode.

5. Produce an exponent based on the order of magnitude of the integer.

This sequence is a few integer operations, all of which are simple hardware units. The operation is not quite this simple, but it is close. Note that if we have an integer that needs more than p bits, we have to chop off the tail bits of that integer when we convert to float. This results in an inexact conversion, which is signaled by an inexact exception, and rounding as normal. We have a floating point result, so there is no specialization of rounding. Finally, if the integer was zero, we set the result to the positive floating-point zero.

The ranges of integers that can be perfectly converted to floating point and operate exactly like integers is shown in Table 6.1. These are relatively wide ranges, covering most of the bits used to store the format. In this range, addition, subtraction, and multiplication of integers represented in floating point are also exact, so the floating-point formats can pretend to be integers within these ranges.

TABLE 6.1
Numeric ranges in which floating-point numbers can convert perfectly from integers and operate like integers. These ranges correspond to the precision of the numeric format.

Numeric Format	Precision	Perfect Integer Range Min	Max
Half precision	11	-2^{11}	2^{11}
Single precision	24	-2^{24}	2^{24}
Double precision	53	-2^{53}	2^{53}
Quad precision	112	-2^{112}	2^{112}

Outside of the ranges shown in Table 6.1, the floating-point numbers can still represent integer values, but may not be exact. For example, the integer $2^{63} - 1$ can be represented exactly in a 64-bit integer format, but it cannot be represented exactly in double-precision floating point. In this regime, one ULP has a numeric value of 2^{10}, so this number has to be rounded to 2^{63} (or $2^{63} - 2^{10}$). The upside of floating point, however, is its dynamic range, so 2^{63} can be represented even in single precision and stored in a 32-bit number.

6.4 Float to Integer Conversion

Float to integer conversion is similar to the process of rounding to integers, but the final target of the result is not a floating-point number, but an integer. However, there is also one wrinkle in the process because the wide dynamic range of floating point means that integers that would need many more than 64 bits can be stored in floating point. Thus, float to integer conversion must contend with overflow as well as rounding. However, it is still a very fast operation.

Like the process of rounding floating-point numbers to the nearest integer, the rounding mode used by a conversion from floating point to integer is explicit and can be unrelated to the rounding mode used for inexact floating-point operations.

In general terms, the steps of this operation are the following:

1. Determine a shift based on the exponent.

2. Remove the exponent and sign bit from the number and add the implied one bit to the mantissa.

3. Shift by the determined shift factor (left or right).

4. If we shifted right and lost bits, round appropriately and signal inexact. If we shifted left and overflowed, instead return either `MIN_INT` or `MAX_INT` based on the sign, and signal inexact and overflow exceptions.

5. Negate the number appropriately if the sign bit was set.

Step 4 of this process contains the exceptional cases. The floating-point number 1.2_F will need to round to integer 1 (in most rounding modes), which will be an inexact conversion. Similarly $(2^{100})_F$ is a valid single-precision floating-point number, but cannot be represented even as a 64-bit unsigned integer. Thus, we will signal overflow and return the maximum possible integer in the destination format. Another interesting case here is that when converting -10_F to an unsigned integer format, we also have an overflow. In this case, we indicate the overflow with the overflow and inexact exceptions, but also return zero, the minimum unsigned integer.

The exception behavior here matches the rounding operations, signaling inexact when we lose precision. An alternate specified form of these operations does inexact conversions silently, suppressing the inexact exception when the floating-point operand is not an integer. However, the default version of integer conversion implemented in processors does signal an exception[3].

There is also no integer `NaN` or infinity, at least in any sane integer format, so the conversion of a floating-point NaN or an infinity to an integer is undefined, and is considered an invalid operation that raises an invalid operation exception. The infinities are not necessarily treated as large numbers in this case—programming languages and processors are not required to do this although some do. Similarly, `NaN` can produce any result that the designers of the programming language and hardware platform decide. If you cannot check your exceptions but risk sending a NaN into an integer conversion, it's a good idea to have a check for `NaN` beforehand to make sure that nothing has gone wrong.

6.5 String to Float

Conversion of strings to floating-point numbers is conceptually simple, but practically very tricky. This is two conversions in one: Reading a decimal number in ASCII (or UTF-8/16) and then turning that decimal number into a binary floating-point number. Sometimes, those numbers are also in decimal scientific notation. Unlike integers, strings of digits can represent a lot of different numbers, including many numbers that cannot be represented in floating point. Thus, this conversion can overflow or underflow, and can easily be inexact. Thankfully, the precision limit on each floating point format dictates a limit on how many significant digits need to be read to determine the mantissa, but there may be an arbitrary number of digits that are required to determine the order of magnitude.

All of the following strings are valid to convert to floating point, and will produce normal numbers:

`"808562976467199983616"`   `"4.5893e-20"`   `"-.0000000000000555"`

The first example is an integer about 40 times larger than 2^{64}, but is exact in double-precision floating point. A conversion of this number through an integer of 64 bits will fail despite the conversion being exact. The second example is a very small number in scientific notation. This is an alternate valid format for floating-point numbers, and is often preferred when they are small or large. The final number has a large number of leading zeros after the decimal place. The number itself has three significant figures, but is 18 characters long. Neither of the latter two examples here is an exact conversion, and as we will discuss in Section 6.7, there are relatively few decimal numbers with few significant digits that are exactly representable.

String conversions will give you the nearest floating point number to the string that was converted, and gracefully overflows and underflows. Overflow and underflow behave like they would for any arithmetic operation, with numbers that are too small to represent getting flushed to zero (or a nearby subnormal number) and numbers that are too large either becoming infinity or the maximum representable normal number depending on rounding mode. Additionally, a few numbers have special string representations that can be converted to and from strings:

`"NaN"`   `"Infinity"`   `"-0"`

These all convert to the expected numbers, NaN, infinity, and negative zero respectively. The shortened string `"inf"` can also convert to infinity. The conversion of `"NaN"` defaults to a `qNaN` either of a canonical form for the language or with a zero payload. Strings like `"sNaN"` or `"snan"` (usually case-insensitive) can produce a signaling NaN if supported by the language. Any invalid input string will result in a NaN and an invalid exception.

Exceptions on string conversions are specified to be the same as exceptions on other operations. However, since these operations are usually done in software, programming languages that suppress exceptions will often not produce an exception, especially for conversions that are merely inexact.

This process also occurs in compilers. When you specify a floating-point literal in your code, your compiler will convert that into binary at compile time using a string-to-float conversion routine. This allows you to specify any literal you want even though the specified numbers may not be able to be represented exactly in floating point. This is a lesson that many developers learn the hard way, when things are just slightly off (see Section 5.8 for the example of 0.04). Inexactness of these specific conversions can have a tendency to be sticky and noticeable because they induce systematic error compared to the expected behavior. Rounding usually introduces some jitter since some runs of a function round up and some round down, but this is a time when every iteration of a function has the same type of error.

Combining the complexity of this operation with language-to-language variability in the encoding of characters and text strings, floating point conversions from text is always done in software, and the code for this is one of the core parts of the language's standard library. Unsurprisingly, this means that conversion from strings is far slower than conversion from integers, and it is also significantly slower than conversion of strings to integers. The space of strings that can represent floating-point numbers is much larger than the space of strings that represent integers, and this adds complexity on conversions both to and from floating point.

6.6 Float to String

Converting floating-point numbers to strings is symmetrical to conversion of strings to floating-point numbers. In this case, we are encoding a binary floating point number as a decimal string.

Conversion of floating point numbers to strings is often very customizable, with options to force the use of scientific notation or not and to control the number of digits before and after the decimal place. In C and C++, formatting floating point numbers using format strings uses the %f and %e (or %E) formatting directives, with f denoting the automatic conversion and e or E forcing scientific notation (the lowercase gives a lowercase e as the indicator for the exponent, while the uppercase gives an uppercase E). This format string can be further augmented with character counts:

$$\%x.yf$$

where x sets the total number of characters to use and y indicates the number of characters after the decimal place. If x has a leading zero, the formatted output will pad the left of the string with zeros to fill x characters [16]. Many languages have similar formatting options, but the default is to use the fewest characters necessary.

Since floating-point numbers can be encoded in scientific notation, float to string conversion routines will often pick whether to use scientific notation or not based on the exponent of the floating point number. Exponents that are very large or very small will use scientific notation, while exponents that are in the middle of the range do not unless the user specifies that they would prefer scientific notation. In many languages, it is possible to force the use of scientific notation for these numbers and vice versa, but the numbers that default to scientific notation are usually very long when written out. Small numbers have a lot of leading zero digits after the decimal place, and large numbers contain a lot of nonzero digits, but those digits have no information content. Large floating point numbers are made of a large integer times a large power of two, so the trailing digits are completely determined by that power of two—they are not significant digits.

A general rule of floating-point string conversions is that when converted to a string and then converted back, you will get the same floating-point number back. Many systems will work in floating point, then transfer that value to a string to hand to another program that works in floating point, so serialization through strings should not induce numerical error. This is exactly what happens when services communicate using text-based formats, like command line pipes, JSON, or batch-processing of logs. Numerically, the string conversion is transparent unless you specify fewer decimal places than the precision of the format demands.

Floating-point NaNs do not preserve this relationship. A NaN can hold a payload of several bits, but the three-character string representation `"nan"` does not preserve the payload of the NaN, and does not preserve whether that was a signaling NaN.

However, conversion of floating-point numbers to strings offers one interesting trick. Unless otherwise specified, the conversion reduces the precision as much as possible while still *uniquely identifying* a floating-point number. It does not use the fewest number of characters that exactly represent the number, but only enough to disambiguate that number from any other floating-point numbers. As a result, despite that 0.1 is not a valid floating-point number (it is an infinitely repeating decimal), `"0.1"` is a valid string that can come out of a float to string conversion. The exact decimal value of the single-precision number that converts to the string `"0.1"` is:

$$\texttt{toString}\,(0.100000001490116119384765625_F) = \texttt{"0.1"}$$

Similarly, the string `"0.1"` converts to that number, as it is the closest floating point number to the real number 0.1.

As a consequence of this sort of conversion, when you widen a number in a narrower floating-point format to a wider floating-point format, it will often not convert to the same string that you get when you convert it directly from the narrow format. Looking at the example above:

$$\texttt{stringToFloat("0.1")} \overset{IX}{\Rightarrow} 0.100000001490116119384765625_F$$

$$\texttt{toString}(0.100000001490116119384765625_D) \Rightarrow \texttt{"0.10000000149011612"}$$

Like string to float conversions, float to string conversions are always done in software libraries, and are relatively computationally expensive compared to math operations. However, it is still cheaper to produce good logs than to have bad bugs.

6.7   Exactness of Decimals in Floating Point

Many base-ten decimal numbers cannot be represented exactly in floating point. This is relatively easy to see if we look at the fractions that represent these numbers. The following fractions are examples of numbers that have finite-length decimals in binary:

$$\frac{3}{4} \qquad \frac{29}{128} \qquad \frac{456\,773}{262\,144}$$

There are many numbers that have infinite numbers of digits behind the decimal place in both base ten and base two. A few examples are:

$$\frac{1}{3} \qquad \frac{15}{17} \qquad \frac{3}{28}$$

Finally, here are a few examples of fractions that have finite-length decimal expansions in base ten but not in binary:

$$\frac{1}{5} \qquad \frac{15}{25} \qquad \frac{1}{1\,000}$$

The common thread of these numbers is that the numbers that are finite decimals in base two have a denominator that is purely a power of two. Finite decimals in base ten can have prime factors in their denominators of five or of two. This is because the prime factors of ten are five and two. Any number whose denominator in reduced rational form has a prime factor that is not a prime factor of the base cannot be represented with a finite number of decimal places. This is easy to see for binary, since each power of two represents a separate bit position. This means that there is a very limited set of decimal numbers (and a very limited set of fractions) that can be represented in a finite-length binary number.

The set of valid floating-point numbers is even smaller, however. Floating point has a precision limit, so any number that requires more bits than can fit in the mantissa is also out of contention. Many fractions with a power of two denominator (including all of the examples above) are valid, but some are not. This number must be rounded in single-precision floating point:

$$\frac{44\,040\,193}{65\,536} = 10\,1010\,0000.0000000000000001_b$$

The decimal representation of this fraction needs 26 bits of precision, and while floating point gives us precision that is independent of the denominator, single-precision floating point only gives us 24 bits of precision, so the trailing one bit will be rounded despite this being a finite-length decimal in binary.

There is a famous example of a set of string conversions producing a result that looks weird to those who are not familiar with floating point, which comes from the following:

$$\texttt{toStr}\left[\texttt{toDouble("0.1")} + \texttt{toDouble("0.2")}\right] \Rightarrow \texttt{"0.30000000000000004"}$$

You might guess that several inexact exceptions are involved in this chain, and there are four. First, 0.1 and 0.2 are both converted inexactly from strings to floats. Both numbers, when put in rational form, have a denominator that is a multiple of five. The conversions are:

$$\texttt{stringToDouble("0.1")} \overset{\text{IX}}{\Rightarrow} 1.100110011001\ldots1010_b\texttt{D}(-4)$$

$$\texttt{stringToDouble("0.2")} \overset{\text{IX}}{\Rightarrow} 1.100110011001\ldots1010_b\texttt{D}(-3)$$

0.1 and 0.2 are not exactly representable in binary floating point, so we have to round to the nearest values. In this case, it is a pattern of 1001's repeating 12 times before a final 1010, with an exponent of -4 for 0.1 and -3 for 0.2. These two numbers are slightly greater than 0.1 and 0.2 respectively, and the conversion is inexact. The sum of these two numbers is:

$$1.1001\ldots1010_b\texttt{D}(-4) + 1.1001\ldots1010_b\texttt{D}(-3) \overset{\text{IX}}{\Rightarrow} 1.0011\ldots0100\texttt{D}(-2)$$

Due to the trailing one bits and the different order of magnitude of the result, our third inexact exception is in this addition. The rounded result is slightly above 0.3, as expected, but its error is double that of the error of the conversion results. We have gotten unlucky here, and all three operations rounded up. Due to that rounding, our result is not the closest double-precision number to 0.3:

$$1.0011\ldots0011\texttt{D}(-2) = 0.2999999999999999888978\ldots \approx 0.3 - 1.11 * 10^{-17}$$

$$1.0011\ldots0100\texttt{D}(-2) = 0.3000000000000000444089\ldots \approx 0.3 + 4.44 * 10^{-17}$$

so $1.0011\ldots0011\texttt{D}(-2)$ is the canonical double-precision form for the string "0.3". When we convert our result to a string, the conversion back to a string is done with enough precision to uniquely identify it, so it is "0.30000000000000004" to distinguish it from "0.3". For good measure, this conversion to a string is also inexact. Four inexact operations create an error that is noticeable to the user. This is a large part of why floating-point numbers are not suitable for calculator applications, which often use arbitrary-precision formats, and must be used with care in monetary calculations—and often several explicit rounding operations.

Significance Tracking in Floating Point

As a consequence of this inaccuracy, many decimals that have few significant digits will have inexact representations in floating point. When considering numbers like 0.1 or 0.7 which have only one digit behind the decimal place, the only number that is exact in floating point is 0.5. Similarly, with two significant digits, only 0.25, 0.50, and 0.75, are exact when converted to and from floating point. If you are thinking about precision of calculations in terms of decimal significant digits, the floating point results will often surprise you. If you can instead work with an idea of *significant bits*, transferring significance tracking to base two from base ten, floating point arithmetic will offer fewer surprises and fewer inexact calculations.

Given a specified margin of error, ϵ, a number will have $\log_{10}(\epsilon)$ significant digits, and $\log_2(\epsilon)$ significant bits. These numbers are often rounded up. Thus a conversion of significant digit count to significant bit count is a straight multiplication by $\log_{10}(2) \approx 0.3$ although rounding twice to whole numbers can give you some error on the number of significant bits you have.

Preventing inexact calculations is now a matter of counting guaranteed trailing zero bits. When you convert numbers with k significant bits to a floating-point format with precision p, the result will have at least $p-k$ trailing zeros. From there, multiplications will reduce the number of trailing zeros by k, and additions of numbers of similar magnitude will reduce the number of trailing zeros by a few bits. You only risk inexact calculations when you reach zero trailing zeros. Additions of numbers that vary greatly in magnitude will cut into this budget significantly, however.

Significance tracking in terms of significant bits and significant digits will give you results of decent accuracy and precision as long as you take the step to round the calculation to the appropriate number of digits. In base b, you can round to k decimal places with the operation:

$$\texttt{roundToDecimalPlaces}_b(x, k) = \frac{\texttt{round}(x \times b^k)}{b^k}$$

The example above did not have such a rounding step, which induced a visible amount of error. With base-ten significant digits, the division will be inexact, but with base-two significant bits, the division will always be exact.

6.8 Bit Casting Floating-Point Numbers

An interesting way to get in and out of floating point is to do bit casting. Instead of converting to an integer, bit casting will get you an integer with the exact same bits as the floating point number. The value of that integer will be very different than the value of the floating-point number, but the internal representation will be the same.

Bit casting is how you can do bitwise manipulation on the floating point number, and is not a standard operation, but a consequence of how floating-point numbers are stored. It allows you to use integer operations to construct floating-point numbers out of their parts and to isolate individual pieces of floating-point numbers. Several languages have a backdoor method through which this can be done, even though setting an integer equal to a floating-point number (and vice versa) usually does a conversion.

Bit casting floating-point numbers is relatively rare compared to conversion, and usually done by people who are up to something. However, bit casting floating point numbers is a unique and powerful mathematical operation that we will take full advantage of in Section 10.6 to create incredibly fast approximations of transcendental functions, and many of the problems in earlier chapters require bit casting to construct and operate on floating-point numbers in their constituent parts.

Check Your Understanding

Problem 6.1. Find the smallest 64-bit integer that has error greater than $1\,000$ when converting to double-precision floating point. Find the smallest double-precision floating-point number greater than $1\,000$ that has an error of any kind when converting to 64-bit integer.

Problem 6.2. Write a function that converts from a rational number, $\frac{n}{d}$, represented as a pair of 64-bit integers to a double-precision floating-point number. Ensure that your function can convert $\frac{2^{64}-1}{6\,700\,417}$ with correct rounding.

Problem 6.3. Find the shortest string (in character count) that overflows a conversion to double-precision floating point. Find the shortest string that underflows a conversion to double-precision floating point. Finally, find the shortest string that converts exactly to a double-precision number, and find the shortest string that converts inexactly.

Problem 6.4. Write a function that efficiently converts decimal strings to single-precision floating point (disregarding scientific notation) with the simplifying assumption that each string will only ever be 10 characters long or less. Use any integer or floating-point operations you need (including double-precision operations), except for string conversions. Use the built-in conversion operation as a reference for testing. Run benchmarks and see if there are any sets of numbers where your version is faster than the built-in function.

7

Numerical Operations

We discussed before that floating point is a format designed for numerical computing. Now is its time to shine. Math libraries come with a number of numerical operations that are built as correctly as possible, sometimes bit-accurate. These functions usually have efficient implementations that allow you to stand on the shoulders of giants for most of your numerical calculations.

A number of these operations are either combinations of operations from earlier chapters or they are transcendental functions that are implemented for you in languages and math libraries. A deeper dive into how transcendental functions are implemented is in Chapter 10, but in this chapter we will be discussing these operators from the perspective of a user.

7.1 Remainder and Modulus Operations

In order to emulate the behavior of the integers, languages and the floating point standard set definitions for floating-point remainder and modulus operators. However, the behavior of these operators varies language-to-language, especially when languages intend to emulate integer arithmetic using floating-point values.

The standard floating-point remainder operation is not the same as the integer modulus of a number. Instead of computing the remainder of a truncated division (the behavior of integer division), it computes the remainder of a rounded division. Mathematically, the remainder operation is equal to:

$$\texttt{remainder}(x, y) = x - y \times \texttt{roundToIntegerNearest}\left(\tfrac{x}{y}\right)$$

Note that rounding to nearest (with ties to even) is used, not rounding toward zero you would expect for an integer remainder calculation. This gives the smallest magnitude remainder that is possible, but it means that the remainder can be negative when both x and y are positive:

$$\texttt{remainder}(5.0_\mathrm{F}, 3.0_\mathrm{F}) = 5.0_\mathrm{F} - 3.0_\mathrm{F} \times \texttt{round}\left(\frac{5.0_\mathrm{F}}{3.0_\mathrm{F}}\right) = 5.0_\mathrm{F} - 3.0_\mathrm{F} \times 2.0_\mathrm{F} \Rightarrow -1.0_\mathrm{F}$$

DOI: 10.1201/9781003565543-7

If you are familiar with the normal remainder operation, you would expect the remainder's sign to match the sign of x, the dividend. That is not the case for this operation.

Languages like C and JavaScript provide a remainder similar to the remainder from integer division. In the case of C, this operator is in addition to the standard floating-point remainder, but in the case of JavaScript, it is in its place. This operation is defined by:

$$\texttt{fmod}(x, y) = x - y \times \texttt{roundToIntegerTowardZero}\left(\frac{x}{y}\right)$$

Note that while the C standard library calls this function "**fmod**", it is not taking a modulus in the mathematical definition of the word, but a remainder [16]. This is also the behavior of JavaScript's remainder operator [8]. This operation can be constructed by using the **remainder** function, and then correcting the sign to match the sign of x then adding y if the remainder's sign is not equal to the sign of x. However, the floating-point **remainder** operation cannot easily be constructed from **fmod**.

In both cases of remainder and modulus, it is invalid to have an infinite x or a zero y. The special cases are shown in Table 7.1. Additionally, if the remainder would be zero, its sign is the same sign as x. Remainders are always exact by definition.

A few languages like Python provide a modulo operator instead of a remainder. While the remainder operator follows the sign of x, a modulo operator follows the sign of y. The behavior only differs when x and y have different signs. A modulus can be constructed out of remainder (**fmod**) functions by the following formula:

$$\texttt{modulus}(x, y) = \texttt{fmod}\left[\texttt{fmod}(x, y) + y, y\right]$$

These operations allow you to construct the last remaining piece of integer arithmetic that is not normally supported by floating point. They also have

TABLE 7.1
Special cases of remainder operations. Cases are the same for **remainder** and **fmod**. In this table, a represents any number (non-NaN).

| Operands | | Result | |
x	y	remainder(x, y)	Exceptions
a	$\pm\infty$	x	
± 0.0	a	± 0.0	
a	± 0.0	NaN	IO
$\pm\infty$	a	NaN	IO
Any	NaN	NaN	IO for sNaN
NaN	Any	NaN	IO for sNaN

niche applications in numerical computing. The standard version is the most flexible and extensible remainder, using rounding to nearest, but versions more aligned with integer arithmetic are often provided by programming languages.

7.2   Exponentials and Logarithms

Exponentials and logarithms are some of the most common transcendental functions to see in scientific computing, and as such, they are part of every math library. These basic functions are often provided in base two, base ten, and base e, although the internal implementation inside many math libraries is in base two, with the other bases covered by a strategically placed constant multiplication. The common exponential functions are exp, exp2, and exp10, while their logarithm counterparts are log, log2, and log10.

Many math libraries also have functions that compute exponential functions minus one. For example, the function exp2m1:

$$\texttt{exp2m1}(x) = 2^x - 1$$

Symmetric to these functions are the logarithms of $x+1$, which are also often found in math libraries:

$$\texttt{log10p1}(x) = \log_{10}(x+1)$$

In some math libraries, the expm1 functions can get special handling, as the exponentials minus one will produce results close to zero at different places than the exponential functions will, and the results near zero need better absolute error to have a fixed number of ULPs of error due to the density of numbers near zero. Thus, library implementers may prefer to use different approximations to get the best relative accuracy in this region. The same goes for the logp1 functions. Not all standard libraries have these functions, and many programmers will blissfully use $(\exp(x) - 1)$ instead of $\texttt{expm1}(x)$.

In base two, exponential and logarithm can be constructed by separating the parts of the number, using a similar idea to how division and square root were computed in Chapter 4. Taking the example of logarithm, we note that the sign bit does not factor into the calculation since we are restricted to positive x (logarithm is undefined for negative x). Then, splitting x into its exponent and mantissa:

$$\texttt{log2}(x) = \texttt{log2}\left[2^{e_x}(1.m_x)\right] = \texttt{toFloat}(e_x) + \texttt{approxLog2}(1.m_x)$$

This means that a log in base two is given to us by taking the exponent of the number and then adding that to an approximation of the logarithm of the significand. Note that the change-of-base formula for logarithms then allows us to extend from base two to any base:

$$\log_b(x) = \frac{\log_2(x)}{\log_2(b)}$$

Some math libraries will give separate implementations for the best bit accuracy instead of using this formula, but many will use this formula, as it only costs a small amount of accuracy in the worst case.

Unlike division and square root, where the parts of the number get computed completely separately, the logarithm formula above adds the separated parts of the floating-point number to get one floating-point result. Exponential functions are symmetric to logarithms, generating the parts of the number separately based on the entire floating-point input. First, we take the integer part of the x input to construct the exponent of the result, then get the exponent by approximating based on the fractional part.

While exponentials can overflow and underflow, they are relatively well-behaved in terms of exceptional conditions, only having exceptions when a NaN is passed in. At the limits, $\text{exp2}(\infty) \Rightarrow \infty$ as expected, and $\text{exp2}(-\infty) \Rightarrow +0.0$. Logarithms have a slightly more interesting set of exceptional cases, including the first case of "division by zero" that does not involve dividing by zero:

$$\log(0.0_\text{F}) \overset{\text{DZ}}{\Rightarrow} -\infty$$

Since an operation on a number produces an infinity, we indicate division by zero for the case of logarithms of zero. Additionally, logarithm is undefined for any number less than zero, so these cases are all invalid:

$$\log(x < 0) \overset{\text{IO}}{\Rightarrow} \text{NaN}$$

These cases are uniform for all of the logarithm operations, and shifted over for the log-plus-one operations.

7.3   Powers and Roots

Extending exponentials into two dimensions, we get the power functions used to construct arbitrary powers of numbers. These are some of the most flexible and powerful functions for numerical computing. The power function computes the function:

$$\text{pow}(x, y) = x^y$$

This allows us to compute arbitrary exponentials, arbitrary powers of numbers, and arbitrary roots. The downside of this function is that it is comparatively slow and hard to make perfectly accurate when x and y are "weird" numbers. Similarly, the flexibility of this function means that the floating-point version has several compromises.

Internally, one way that this function can be computed, and an outline for the intended implementation for `pow` is the following:

$$x^y = \left[2^{\log_2(x)}\right]^y = 2^{y\log_2(x)}$$

This identity only works if x is positive, however, which means that most implementations will have a special case for negative x and integer y, but will yield NaN for any negative x and non-integer y, indicating an invalid operation. Many of these results will have complex results either way, but some of them are unintuitive:

$$\text{pow}\left(-8.0_F, \tfrac{1}{3}_F\right) \overset{\text{IO}}{\Rightarrow} \text{NaN}$$

The cube root of -8 is -2, but the `pow` function can't be used to compute it for you because $\tfrac{1}{3}$ is inexact and so the true result of the calculation above is actually a complex number. The cases for zero and infinity take some thought, but all follow a logical pattern, shown in Table 7.2. One to any power is one, even when that power is a qNaN, and anything to the zeroth power is also one. From there, we have several ways to construct the equivalent of $\tfrac{1}{0}$ and $\tfrac{1}{\infty}$ using `pow`, triggering a divide by zero exception when we construct $\tfrac{1}{0}$ out of two finite operands (compare the third and fifth rows of Table 7.2). For values of x where x^y grows as y grows, x^∞ is infinite, while values of x that shrink as y grows yield zero. Finally, the domain restriction around negative x values is loosened for zero-valued and infinite-valued x.

The need to compute both an exponential and a logarithm means that `pow` is often one of the most inaccurate functions in any math library. It is also guaranteed to be computed in software, and has one of the longest computation times of any function in a math library.

Narrowing the function space somewhat, when you have an integer n where $y = n$ or $y = \tfrac{1}{n}$, there are a few other functions that may be available and can be faster: The `pown` and `rootn` functions. Telling the implementation that you are taking the nth power or nth root allows math libraries to compute these functions much more quickly and more accurately than the generic power function.

$$\text{pown}(x, n) = x^n \qquad\qquad \text{integer } n$$
$$\text{rootn}(x, n) = \sqrt[n]{x} \qquad\qquad \text{integer } n$$

In this case, instead of approximating log, `pown` and `rootn` can be done with algorithms that are more restricted and more accurate than `pow`. These functions also sidestep the issues with negative values of x that show up with the `pow` function, since the space of n is limited enough that it is possible to know that the user wants a real-valued nth root rather than an arbitrary power. These functions are not part of the C math library, but are in the standard libraries of some other programming languages like C#.

TABLE 7.2
Special cases of the power (`pow`) function. In this table, O represents an odd positive integer. Note that NaN inputs can produce non-NaN results, and asymptotic behavior depends on evenness and oddness.

Operands		Result	
x	y	$\text{pow}(x, y)$	Exceptions
Default One Outputs			
Any including qNaN	± 0.0	1.0	
1.0	Any including qNaN	1.0	
Zero-valued x			
± 0.0	$-O$	$\pm \infty$	DZ
± 0.0	O	± 0.0	
± 0.0	$-\infty$	∞	
± 0.0	∞	$+0.0$	
± 0.0	$y > 0$ not odd integer	$+0.0$	
± 0.0	$y < 0$ not odd integer	∞	DZ
Infinite x			
$\pm \infty$	∞	∞	
$\pm \infty$	$-\infty$	$+0.0$	
∞	$y < 0$	$+0.0$	
∞	$y > 0$	∞	
$-\infty$	$-O$	-0.0	
$-\infty$	O	$-\infty$	
$-\infty$	$y > 0$ not odd integer	∞	
$-\infty$	$y < 0$ not odd integer	$+0.0$	
Infinite y			
$-1 < x < 1$	$-\infty$	∞	
$x < -1$ or $x > 1$	$-\infty$	$+0.0$	
$-1 < x < 1$	∞	$+0.0$	
$x < -1$ or $x > 1$	∞	∞	
-1.0	$\pm \infty$	1.0	
Invalid Operations and NaNs			
$x < 0$	non-integer y	NaN	IO
Any $\neq 1.0$	qNaN	NaN	
qNaN	Any $\neq 0.0$	NaN	
Any	sNaN	NaN	IO
sNaN	Any	NaN	IO

TABLE 7.3

Special cases of the integer power (`pown`) function. In this table, O represents an odd positive integer and E represents an even positive integer.

Operands		Result	
x	n	$\text{pown}(x, n)$	**Exceptions**
Any except `sNaN`	0	1.0	
± 0.0	$-E$	∞	DZ
± 0.0	$-O$	$\pm\infty$	DZ
± 0.0	E	$+0.0$	
± 0.0	O	± 0.0	
∞	$n > 0$	∞	
∞	$n < 0$	$+0.0$	
$-\infty$	O	$-\infty$	
$-\infty$	E	∞	
$-\infty$	$-O$	-0.0	
$-\infty$	$-E$	$+0.0$	
`qNaN`	Any $\neq 0$	`NaN`	
`sNaN`	Any	`NaN`	IO

The special cases of **pown** and **rootn** are found in Tables 7.3 and 7.4 respectively. Note that the **rootn** function can differ from the square root function in one corner case:

$$\text{rootn}(-0.0_F, 2) \Rightarrow +0.0_F$$

This function actually behaves as you might expect for square root rather than yielding negative zero. Additionally, you cannot take the zeroth root of a number, while the zeroth power of anything but a signaling `NaN` is one. Even roots of negative numbers are invalid since they would be imaginary, while odd roots of negative numbers are valid. It is possible to take negative and positive powers and roots, and the asymptotic behavior toward zero and infinity is rational.

Finally, there is also a specialized cube root function in some math libraries [16]. The cube root function behaves exactly like **rootn**$(x, 3)$, but can be faster knowing that you are strictly taking a cube root. With this restriction, internal methods can be somewhat more optimized to the point where a cube root in software is a similar speed to a square root done in software (although square roots are practically always done in hardware). Knowing the root exactly means that a textbook Newton-Raphson approximation (see Section 10.3) can be used, taking less than 20 floating-point operations from start to finish, which makes one of the fastest operations in this chapter.

While the power function is powerful and flexible, it is also very slow to support that power and flexibility, and many implementations are relatively inaccurate. Using a specialized version of a power or exponential function will

TABLE 7.4

Special cases of the integer root (`rootn`) function. In this table, O represents any odd positive integer and E represents any even positive integer. Note that the even roots of negative x are invalid, while the odd roots are not.

Operands		Result	
x	n	`rootn`(x, n)	**Exceptions**
± 0.0	$-E$	∞	DZ
± 0.0	$-O$	$\pm\infty$	DZ
± 0.0	E	$+0.0$	
± 0.0	O	± 0.0	
∞	$n > 0$	∞	
∞	$n < 0$	$+0.0$	
$-\infty$	O	$-\infty$	
$-\infty$	$-O$	-0.0	
$x < 0$	$\pm E$	NaN	IO
Any	0	NaN	IO
qNaN	Any	NaN	
sNaN	Any	NaN	IO

give you better results faster. However, if you need to compute something like $2.43768^{1.5862}$, the `pow` function is there to help.

7.4   Reciprocal Square Root

Just like square root is a first-class citizen due to graphics calculations, reciprocal square root (also confusingly called "inverse square root") is a very common operation for math libraries and processor hardware to support. Reciprocal square root is useful for operations such as vector normalization:

$$\hat{\mathbf{v}} = \frac{\mathbf{v}}{|\mathbf{v}|} = \left(\frac{v_x}{\sqrt{v_x^2 + v_y^2 + v_z^2}}, \frac{v_y}{\sqrt{v_x^2 + v_y^2 + v_z^2}}, \frac{v_z}{\sqrt{v_x^2 + v_y^2 + v_z^2}} \right)$$

Since floating point division and floating point square root are both relatively slow operations, it makes sense to combine the two into a $\frac{1}{\sqrt{x}}$ operation, the reciprocal square root. We can then use hardware to approximate this result directly instead of using it for two separate operations. One of the main algorithms for fast square root calculation actually involves calculating a reciprocal square root as an intermediate result, so we can take that intermediate directly when we need to compute things like vector norms.

Floating point specifies an optional reciprocal square root operation, although several processors and math libraries do not implement this function directly. The closest that you often get is an approximate reciprocal square root, such as the `VRSQRT14SD` instruction on x86 CPUs, which computes a very fast approximate reciprocal square root that has 14 correct bits. Nvidia GPUs offer an exact reciprocal square root, as would be expected given the usage of the function in graphics calculations, instead of having a square root instruction. A square root is then constructed by computing $x \times \frac{1}{\sqrt{x}}$. In contrast, CPUs will often have a square root operation and give you an inverse by combining that with a division.

Where it is implemented, the reciprocal square root operator is intended to operate exactly like the combination of a square root and a division operation. That implies one weird case for reciprocal square roots:

$$\texttt{rsqrt}(-0.0_\texttt{F}) = \frac{1}{\sqrt{-0.0_\texttt{F}}} \Rightarrow \frac{1}{-0.0_\texttt{F}} \stackrel{\texttt{DZ}}{\Rightarrow} -\infty$$

The most common application of reciprocal square roots is the calculation of distances, though, and the final operation before the square root in $v_x^2 + v_y^2$ is an addition, which will not produce a negative zero unless both operands are negative zero. Since each number is squared, a negative zero will be impossible for either one to produce. In this common case, the absurd possibility of a distance of negative infinity will never occur, even when v_x and v_y are both negative zero.

This operator offers a straightforward speedup when it is available. Compilers such as the CUDA compiler will infer this operation for you. The fast approximate inverse square root is a more blunt tool that you will have to invoke manually when you don't care so much about the error of the operation and prefer to avoid a pair of relatively long operations.

7.5 Trigonometric Functions

The trigonometric functions are the other major class of numerical functions that are common to see in simulations and geometric calculations, so floating-point math libraries provide built-in support for trigonometric functions. The trigonometric functions supported fall into the following categories, measuring angle in radians and in pi-radians (half revolutions):

- The basic trigonometric functions: `sin`, `cos`, and `tan`.

- The trigonometric functions with arguments multiplied by pi: `sinPi`, `cosPi`, and `tanPi`.

- The inverse trigonometric functions: `asin`, `acos`, and `atan`.

- The inverse trigonometric functions with outputs multiplied by pi: `asinPi`, `acosPi`, and `atanPi`.

- A special form of arctangent with both x and y specified: `atan2` and `atan2Pi`.

These functions actually use two different units for angles, with an irrational relationship between them. The normal functions measure angles in radians, where a full circle is $x = 2\pi$ radians. The "Pi" functions compute the same things, but measure angles in half revolutions of a circle, where a full circle is $x = 2$ half-revolutions. The Pi functions have the advantage that large floating-point numbers are all integers, so for large inputs, instead of trying to figure out a position around the circle to compute sines and cosines, the Pi functions are trivial. For example, compare cosine to `cosPi` with large x:

$$\cos(100\,0000) \overset{\text{IX}}{\Rightarrow} 0.936752 \qquad \texttt{cosPi}(100\,0000) \Rightarrow 1$$

$$\cos(10\,0000\,0000) \overset{\text{IX}}{\Rightarrow} 0.837887 \qquad \texttt{cosPi}(10\,0000\,0000) \Rightarrow 1$$

$$\cos(1\,0000\,0000\,0000) \overset{\text{IX}}{\Rightarrow} 0.791446 \qquad \texttt{cosPi}(1\,0000\,0000\,0000) \Rightarrow 1$$

The large integers on the left side are some fraction of a circle, so we have to figure out where we are to get a sine or cosine. When the precision of numbers is so rough that one ULP exceeds 2π, we don't have enough precision to decide where we are on a circle at all. Using the Pi functions lets us simply set the result. It is also easier to convert degrees into half-revolutions than into radians while maintaining accuracy, since the ratio of 180 degrees per half-revolution can be represented exactly in floating point, while $\frac{180}{\pi}$ cannot. By their nature, the trigonometric functions in radians are almost always inexact, while the trigonometric functions in half-revolutions can be exact more often.

Additionally, the `tan` function almost never returns an infinite value because π is irrational, so its value can only get arbitrarily large, while the `tanPi` function reaches a value of positive or negative infinity at many floating-point values, lying halfway between the integers (e.g., `tanPi`$(0.5) = \infty$), indicating the infinite value with a divide by zero exception. The tangent function can overflow at input values near $x = \left(n + \frac{1}{2}\right)\pi$, but since all of these x values are irrational, it will never yield a division by zero exception since these overflows are inexact, not exact. For all positive integers n, the `tanPi` function follows the pattern:

$$\texttt{tanPi}\left(2n + \tfrac{1}{2}\right) \overset{\text{DZ}}{\Rightarrow} \infty \qquad\qquad \texttt{tanPi}\left(-2n - \tfrac{1}{2}\right) \overset{\text{DZ}}{\Rightarrow} -\infty$$

$$\texttt{tanPi}\left(2n + 1\right) \Rightarrow -0.0 \qquad\qquad \texttt{tanPi}\left(-2n - 1\right) \Rightarrow +0.0$$

$$\texttt{tanPi}\left(2n + \tfrac{3}{2}\right) \overset{\text{DZ}}{\Rightarrow} -\infty \qquad\qquad \texttt{tanPi}\left(-2n - \tfrac{3}{2}\right) \overset{\text{DZ}}{\Rightarrow} \infty$$

$$\texttt{tanPi}\left(2n + 2\right) \Rightarrow +0.0 \qquad\qquad \texttt{tanPi}\left(-2n - 2\right) \Rightarrow -0.0$$

At zero, the `tanPi` and `tan` functions follow the sign of the zero passed in, and yield negative zero when a negative zero argument is passed in. This makes the `tanPi` and tangent functions odd under most rounding modes:

$$\mathtt{tanPi}(-x) = -\mathtt{tanPi}(x)$$

The `sin` and `sinPi` functions are odd and the `cos` and `cosPi` functions are even. This means that the following relationships hold:

$$\mathtt{sinPi}(-x) = -\mathtt{sinPi}(x)$$
$$\mathtt{cosPi}(-x) = \mathtt{cosPi}(x)$$

This extends to all of the integer values, where $\sin(x) = 0$, and the values halfway between the integers where cosine is zero. For positive integer n (including $n = 0$):

$$\mathtt{sinPi}(n) \Rightarrow +0.0 \qquad \mathtt{cosPi}\left(n + \tfrac{1}{2}\right) \Rightarrow +0.0$$
$$\mathtt{sinPi}(-n) \Rightarrow -0.0 \qquad \mathtt{cosPi}\left(\tfrac{1}{2} - n\right) \Rightarrow +0.0$$

The zero-crossings of `sinPi` on the negative side of the x-axis are all negative zero, while they are positive zero on the positive side, while the zero-crossings of the `cosPi` function are always at positive zero. This is due to the strict oddness and evenness of the functions.

While the trigonometric functions have an infinite input domain and a relatively restricted output range, the `atan` and `acos` functions have a restricted domain under which they produce a non-NaN value. Outside of the domain $[-1, 1]$, these functions will give an invalid operation exception. The domains and ranges of these functions are shown in Figure 7.1. While the ranges of the inverse trigonometric functions are usually bounded by π or $\frac{\pi}{2}$, once rounding comes into play, it is possible for the output of these functions to exceed the limits that might be expected in some rounding modes. This means that you

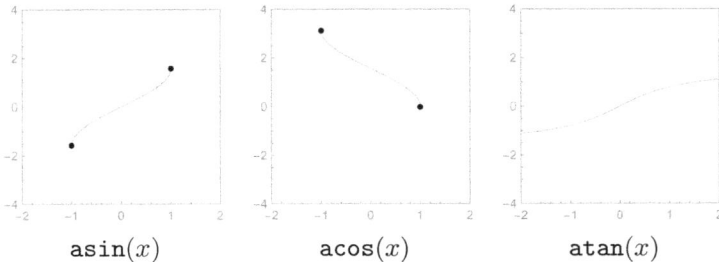

$\mathtt{asin}(x)$ $\qquad$ $\mathtt{acos}(x)$ $\qquad$ $\mathtt{atan}(x)$

FIGURE 7.1
Plots of the inverse trigonometric functions. The arcsin and arccos functions have limited domain. The `asinPi`, `acosPi`, and `atanPi` functions are scaled down vertically by a factor of pi.

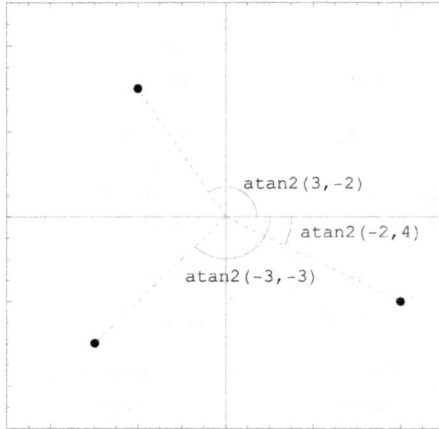

FIGURE 7.2
Examples of the two-argument arctangent. The function measures the angle
from the positive x-axis to a given point.

may need to take care when using the radian-valued functions to check for a
jump from one region to the next—the Pi functions do not have this problem,
as their bounds are exact floating-point numbers. One class of inverse trigonometric function has additional flexibility: The `atan2` and `atan2Pi` functions.
The `atan2` functions give the angle from the positive x-axis to the line from the
origin to the point (x, y) in Cartesian space. This is an extension of the arctan
function that brings back the geometric interpretation of the arctan function.
The arguments to `atan2` are provided "backwards" with the y-coordinate as
the first argument or the side length of the right triangle opposite the angle.

A few graphical examples are shown in Figure 7.2. The output of `atan2` sits
in the interval $[-\pi, \pi]$ (unrounded), and the output of `atan2Pi` is inside $[-1, 1]$.
This is different than the `atan` function, whose range is restricted $\left[-\frac{\pi}{2}, \frac{\pi}{2}\right]$.
Since we have a vector whose angle to measure, we get an extra degree of
freedom in terms of triangles that can be represented by the input. In one
function, this function also gives us a simple transformation from Cartesian
to polar coordinates.

The `atan2` function has a number of special corner-case values, and some
relatively distinct handling of these corner cases. The corner cases of `atan2`
are shown in Table 7.5. None of the relevant corner cases causes overflow
or underflow, but they do cause *inexactness* due to the fact that all of the
non-zero outputs are irrational. The `atan2Pi` function does not have inexact
outputs at these points, because its output is exact. We essentially have to
guess when the inputs are too large or too small to measure.

TABLE 7.5
Special cases of the two-argument arctangent functions. Here, a represents any positive number. `atan2Pi` provides the same outputs as `atan2` scaled down by a factor of π.

Operands		Results		Exceptions	
y	x	`atan2`(x,y)	`atan2Pi`(x,y)	`atan2`(x,y)	`atan2Pi`(x,y)
± 0.0	-0.0	$\pm \pi$	± 1.0	IX	
± 0.0	$+0.0$	± 0.0	± 0.0		
± 0.0	$-a$	$\pm \pi$	± 1.0	IX	
± 0.0	a	± 0.0	± 0.0		
a	± 0.0	$\frac{\pi}{2}$	$\frac{1}{2}$	IX	
$-a$	± 0.0	$-\frac{\pi}{2}$	$-\frac{1}{2}$	IX	
$\pm a$	∞	± 0.0	± 0.0		
$\pm a$	$-\infty$	$\pm \pi$	± 1.0	IX	
∞	$\pm a$	$\frac{\pi}{2}$	$\frac{1}{2}$	IX	
$-\infty$	$\pm a$	$-\frac{\pi}{2}$	$-\frac{1}{2}$	IX	
$\pm \infty$	∞	$\pm \frac{\pi}{4}$	$\pm \frac{1}{4}$	IX	
$\pm \infty$	$-\infty$	$\pm \frac{3\pi}{4}$	$\pm \frac{3}{4}$	IX	
Any	NaN	NaN	NaN	IO for sNan	
NaN	Any	NaN	NaN	IO for sNan	

All of the functions mentioned in this section are difficult to compute, and thus use approximation algorithms in software. The `atan2` function is somewhat more difficult to compute than the rest, but they are all on a similar scale in terms of speed, around the same computation difficulty as an exponential. Math libraries compute each one with good accuracy, although accuracy can suffer as x gets large for the radian-based trigonometric functions.

For trigonometry in floating point, the `Pi` functions are more numerically useful than the basic trigonometric functions because key angles like the 90-degree angle can be represented by exact values in floating point. Unless you are required to work in units of radians, these functions will produce normal behavior and exact results more often than their counterparts. However, all of the trigonometric functions in floating point have radian-valued definitions as well as half-revolution-valued definitions. Finally, there is a special two-argument arctangent function that is useful for measuring angles when given coordinates in space. All of these functions are inexact the vast majority of the time, but they are still useful.

7.6 Hyperbolic Functions

The hyperbolic functions are cousins of the trigonometric functions and exponential functions. These functions are shown in Figure 7.3. All of these functions can be constructed out of exponentials, but math libraries will almost always provide a function that is faster and more accurate than trying to manipulate the exponentials instead.

There are only six floating-point hyperbolic functions, the three basic functions, `sinh`, `cosh`, and `tanh`; and their inverse counterparts, `asinh`, `acosh`, and `atanh`. Each hyperbolic function is defined in terms of exponential functions:

$$\sinh(x) = \frac{\exp(x) - \exp(-x)}{2}$$

$$\cosh(x) = \frac{\exp(x) + \exp(-x)}{2}$$

$$\tanh(x) = \frac{\sinh(x)}{\cosh(x)} = \frac{\exp(x) - \exp(-x)}{\exp(x) + \exp(-x)}$$

We can see that as x moves away from zero, if we evaluate these definitions directly, we will be adding a large number to a small number, which is a recipe for inaccuracy and precision loss. Specifically looking at `tanh`, we get a function that moves between negative and positive one with its exact value depending on the ratio of a pair of large-plus-small additions. Unlike `sinh` and `cosh` which both get large and approach an exponential for large x, allowing us to use the exponential as a perfect approximation past a certain point, the `tanh` function does neither.

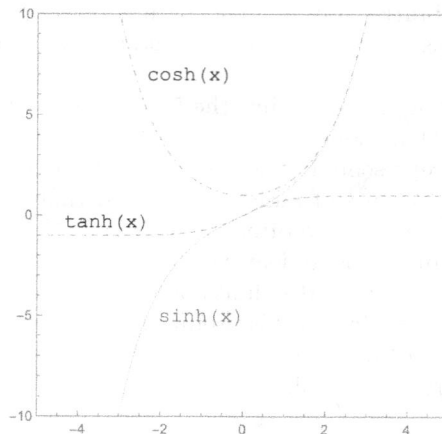

FIGURE 7.3
Plot of the hyperbolic functions.

The hyperbolic functions are often constructed in math libraries with their own approximations rather than relying on the **exp** function, but these functions are harder to approximate than the exponentials, which can be approximated in a separable way, and the trigonometric functions, which have periodic behavior.

Numerically, these functions are generally well-behaved and have infinite domain and range, except for **atanh**, which must have x in the range $[-1, 1]$, and **acosh** which must have x inside $[1, \infty]$. Overflow, underflow, and inexactness will show up as usual, and the **sinh** and **cosh** functions have a relatively small domain where the result is non-infinite due to overflow. The one interesting exceptional case for the hyperbolic functions is the case of **atanh** at the limits:

$$\texttt{atanh}(\pm 1.0) \overset{\text{DZ}}{\Rightarrow} \pm\infty$$

Interestingly, the **asinh**, **acosh**, and **atanh** functions cannot overflow even though they can produce infinite values. In all cases, when x gets large, the magnitude of the result of the function is smaller than x, so we only get an infinite result when x is infinite.

The hyperbolic functions are less common than the other functions from this chapter in geometric calculations and numerical simulations, but the floating-point versions are as fast and accurate as other members of math libraries.

7.7 Statistical Functions

Many math libraries contain functions that relate to statistics and common distributions. Unlike the functions in the rest of this chapter, these functions are not specified in the IEEE 754 standard, and are not implemented uniformly between programming languages. For some, they are in the standard library, and for others, they are the domain of third-party implementations.

The most common statistical function is the error function, $\text{erf}(x)$, and its complement, $\text{erfc}(x)$. These functions are sigmoidal in shape, as shown in Figure 7.4, but are defined as an integral in their real-valued form:

$$\text{erf}(x) = \frac{2}{\sqrt{\pi}} \int_0^x e^{-t^2} dt$$

The complementary function is defined as $\text{erfc}(x) = 1 - \text{erf}(x)$. The error and complementary error functions are common functions in statistics, where they arise from the use of Gaussian distributions, whose cumulative distribution function takes the form of an error function.

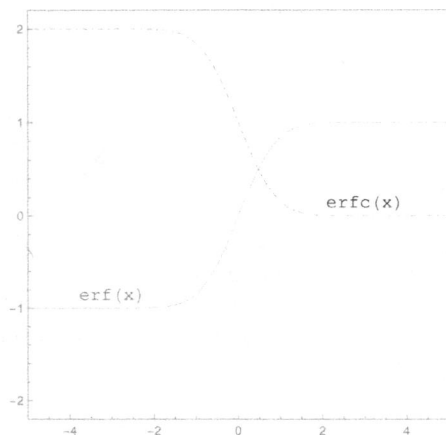

FIGURE 7.4
Plot of the error and complementary error functions.

The error function does not have a simple closed form, which means that evaluation of the exact error function involves taking an integral. However, it is still possible to construct approximations of the error function using polynomials or lookup tables (see Chapter 10). The only region with interesting dynamics is the region near zero, so approximate forms of the error function and similar sigmoid functions are usually relatively small compared to approximations of functions that move across their entire domain. However, this is the first function in this chapter that does not display some form of "nice" behavior—math libraries have to approximate it on its entire relevant domain instead of relying on periodic or fractal behavior.

Numeric formats with higher precision have a wider relevant domain than formats with lower precision, and the `erfc` function needs some care for positive x due to the fact that it approaches zero, the region where the floating-point numbers have high density. When implemented, approximations of the error function in math libraries are usually accurate, but may be slower than some other functions in this chapter that have a smaller relevant domain [20].

A second function that is common to find in math libraries and is also highly relevant to statistics is the gamma function. The gamma function is useful for several statistical tests and distributions, and often uses the moniker "`tgamma`" in math libraries to avoid confusion with a differently defined function previously named "`gamma`". Like the error function, the gamma function is defined by an integral:

$$\Gamma(x) = \int\limits_{0}^{\infty} t^{x-1} e^{-t} dt$$

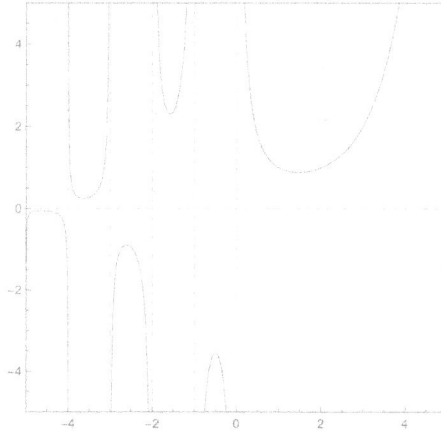

FIGURE 7.5
Plot of the gamma function with its asymptotes shown as dashed lines.

A plot of the gamma function is shown in Figure 7.5. Unlike the error function, the gamma function has significant dynamics across x. Math libraries will also often define the log gamma function (`lgamma`) as a sister to the gamma function, returning the logarithm of $\Gamma(x)$. This function shows up occasionally in fluid dynamics, but is also useful when x is large enough that $\Gamma(x)$ gets very big.

Approximations of the gamma function and its extensions, like the incomplete gamma function, are very difficult to construct. When present in math libraries, the gamma function is usually one of the functions with the most error, sometimes reaching thousands of ULPs, although a few math libraries give a bit-accurate approximation [20]. Most approximations of the gamma function use a convergent series rather than attempting to use one of the methods mentioned later in this book. The two main methods used are Stirling's formula and the Lanczos formula. Corrections to these methods have been proposed for faster convergence, but all involve evaluation of transcendental functions and an infinite series, making precise implementation very difficult [21].

The gamma function also has several points where it is invalid or goes to infinity. The behavior of approximations of the gamma function is not standardized, but the C standard library gives the following error conditions [16]:

- For $\Gamma(\pm 0.0)$, return $\pm\infty$ and raise a `DZ` exception.

- For negative integer x or $x = -\infty$, return `NaN` and raise an `IO` exception.

- For $x = \infty$, return ∞ with no exception.

- Propagate `NaN` as expected.

Math libraries will often contain several other domain-specific functions that are otherwise difficult to approximate, including Bessel functions, beta functions, and the Riemann zeta function. Behavior of these functions is completely defined by the math library used, so while most math libraries will follow the conventions laid out in Chapter 3, there are no guarantees that behavior will be exactly the same from language to language.

Check Your Understanding

Problem 7.1. Find the range of values of x where $\exp(x)$ produces a non-zero, non-infinite result (i.e., where the function does not overflow or underflow).

Problem 7.2. Find the value of x where $\cosh(x)$ is within 1 ULP of $\exp(x)$ in single-precision floating point.

Problem 7.3. Create a speed benchmark to compare $x \times x \times x$ to $\text{pow}(x, 3)$. Which one is faster? Now compare the speed of both to $\text{pow}(x, 3.00001)$.

Problem 7.4. Find the range of values of x where $\text{erfc}(x)$ produces a result that is not equal to 0.0_F or 2.0_F. How many more positive floating point numbers are in this space than negative numbers?

Problem 7.5. Find a value of x in floating point where $\text{tanPi}(x)$ and $\tan(\pi x)$ (using floating-point π) have opposite signs and: (1) $\tan(\pi x)$ is far from zero and (2) $\tan(\pi x)$ is near zero.

Problem 7.6. Compare the output of $\text{pow}(10, x)$ to $\text{exp10}(x)$ by sampling $1\,000\,000$ random points with $-1\,000 \le x \le 1\,000$ (in double precision). What is the maximum difference between the two results? Do the same for $\exp(x)$ and $\text{pow}(e, x)$.

Problem 7.7. Construct a function for $\cosh(x)$ using $\exp$ functions. Compare your function to the math library's implementation of $\cosh(x)$. What is the maximum error between your function and the math library in single-precision floating point?

Problem 7.8. Create a reciprocal function using the square of the result of the fast-reciprocal-square-root assembly instruction on your computer. Measure the maximum error compared to computing $\frac{1}{x}$ directly. Measure the speed compared to the normal approach using division.

8

Bitwise Manipulations

Floating point has a few other operators that work at the level of bit fields, and often don't make as much sense to define mathematically. Some of them, like the absolute value function, are also mathematical operators, but floating point does not treat these functions in the same way it does the numerical and arithmetic operators. Most of these functions don't report exceptions or errors on their inputs, and they can be used in ways that bend or break the floating-point abstraction.

The most prominent example of bending the abstraction is the use of NaN payloads to carry non-floating-point information. This can be helpful for compressing other values by packing them into NaNs or for carrying information about errors. This is enabled by paying attention to the binary encoding of floating point.

8.1   Exponent Manipulation

Some functions allow you to directly manipulate the exponent of floating-point numbers, giving a fast approximate logarithm and a fast scaling function. These functions are essentially just faster ways to do things that could be done with other, slower operators. Practically, these functions are actually short sequences of integer operations with a useful meaning in floating point.

The first of these functions is the `logB` function. Unlike the name suggests, it does not actually compute a logarithm. Instead, `logB`(x) returns the exponent of x as a signed integral value (as though it had infinite exponent range). It can return that result either in floating point or as an integer—like C and C++, we will refer to the integer version as `iLogB`, while the version that returns a floating-point number will be called `logB`. Mathematically, `logB`(x) functions as $\lfloor \log_2(x) \rfloor$ when x is positive. Unlike a logarithm, however, it is possible to take `logB` of a negative number since we are not actually doing a mathematical operation, but extracting the exponent. Pseudocode for `logB` is shown in Algorithm 8.1. The algorithm mostly comprises checks for zero and infinity after extracting the exponent. We also have to give care to subnormal numbers, which need to be scaled up into the normal numbers to correctly get their exponent.

Algorithm 8.1 Pseudocode for `logB`

Input: floating-point x
Output: floating-point $y = \text{logB}(x)$

1:   $i \leftarrow \text{bitCast}_{\text{Int}}(x)$
2:   $e_x \leftarrow (i \gg mBits)$ & $eMask$ ▷ Get exponent bits
3: **if** $e_x = eMask$ **then**   ▷ Infinity or NaN, exponent is all ones
4:      **if** $(x$ & $mMask) = 0$ **then**
5:          **return** ∞
6:      **else**
7:          **return** NaN   ▷ Check for sNaN here is omitted for brevity
8:      **end if**
9: **else if** $e_x = 0$ **then**   ▷ Zero or subnormal number
10:     **if** $(x$ & $mMask) = 0$ **then**   ▷ Zeros
11:        **return** $-\infty$ **with** DZ **exception**
12:     **else**   ▷ Subnormal numbers
13:        $n \leftarrow x \times 2^p$ ▷ Scale up to a normal number
14:        $i_n \leftarrow \text{bitCast}_{\text{Int}}(n)$
15:        $e_n \leftarrow (i_n \gg mBits)$ & $eMask$ ▷ Get exponent
16:        $e_x \leftarrow e_n - p$ ▷ Correct for our scale factor
17:     **end if**
18: **end if**
19: **return** $\text{intToFloat}(e_x + eBias)$ ▷ Unbias exponent and return to float

---

As you can see from the algorithm, the majority of the complexity in `logB` is around handling of special cases. The zeros have an exponent of negative infinity and raise a divide by zero exception similar to the logarithm operation. The infinities have an exponent of positive infinity. Everything else has an exponent that is an integer value. To handle the subnormal numbers, it turns out that the most efficient method on most modern hardware is to simply multiply by the precision of the format to get a normal number and then repeat the procedure of extracting the exponent. It is also possible to replace lines 13–16 with a procedure producing the leading zero count of the mantissa if the multiplication will require a microcode assist or otherwise be slower than the bit manipulation. Also, `logB` cannot overflow or underflow. Its output range, aside from the cases that produce infinities, is relatively limited.

For `iLogB`, we will be producing an integer that does not have infinities or NaN in its output space. Line 19 of the algorithm will simply not return to floating point in this case, but all of the cases that would return an infinity or a NaN will instead return an error result (such as a result that cannot be a valid exponent) and raise an invalid operation exception.

While binary floating point numbers take a base-two logarithm for `logB`, on decimal floating point numbers, `logB` performs the equivalent of a logarithm in base ten. The "B" stands for "base", referring to the base of the format, not "binary".

A complement of `logB` is the `scaleB` function. `scaleB(x, n)` multiplies x by 2^n (actually b^n for floating point in base b), where n must be an integer. Similarly to how `logB` extracts the exponent field, `scaleB` adds to it:

$$\text{scaleB}(x, n) = (-1)^{s_x} 2^{e_x + n}(1.m_x)$$

Even though this is a simple operation, the `scaleB` function also checks for overflow and underflow and indicates the appropriate exceptions if those occur. However, `scaleB` is always exact thanks to the exponent-only nature of the manipulation unless there is an overflow or underflow, or unless the input is a subnormal number. The implementation of `scaleB` looks somewhat similar to `logB`, where most of the code handles special cases. The common path is simpler and faster than a multiplication, but it is fast enough to simply construct a floating point number equal to 2^n through bit manipulation and then multiply it by x, and this has all the correct signaling and error handling built-in.

8.2 Sign Bit Manipulation

Sometimes, all you need to do a useful piece of math is to manipulate the sign bit of a floating-point number. The archetypal example of this is negation, where the sign bit is negated. This flips the number from positive to negative. Notice that this is easier and simpler than it is for integers, where negation can change many of the bits of the number and where the range of positive and negative numbers is different. In floating point, the only difference between a positive number and a negative number of the same magnitude is the sign bit, so to negate, we flip the sign bit. As expected, this also means that negation has no exceptions.

The absolute value operator (`abs` or `fabs`) is another operation done solely on the sign bit. By setting the sign bit to zero, you get the absolute value of a floating point number:

$$|x| = (-1)^0 2^{e_x}(1.m_x)$$

This is another operation that is done without exceptions (even with an `sNaN` input). If you happen to pass a negative NaN into the `abs` function, most languages will give you a positive NaN. Remember that NaNs can carry a sign, even though it is usually meaningless. This operator is actually just a pure manipulation of the sign bit.

The `signBit` operator in some math libraries extracts the sign bit from a floating-point number. The C specification indicates that `signBit(x)` returns "a positive value" (equivalent to a boolean `true`) if x has a negative sign [16] and zero otherwise, while the C++ specification indicates specifically a boolean return type. This is not a standard floating-point function, but is a convenience, and when used as a comparison function, can disambiguate negative zero as being less than positive zero.

A final sign bit manipulation is the `copySign` operation. $copySign(x, y)$ returns the number x with the same sign as y, copying the sign bit from y and the rest of the number from x. If you were wondering in the previous paragraph how to get a negative NaN, the `copySign` operation is usually how they are created. `copySign` is also completely free of exceptions and passes sNaN through. The sign bit manipulation operations, while they have mathematical functions, are bit manipulations at their core, and act as such.

8.3   Iterators

A few floating-point functions act as iterators through the space of floating-point numbers. These functions are the following:

- `nextUp`(x) returns the next floating-point number above x.

- `nextDown`(x) returns the next floating-point number below x.

- `nextAfter`(x, y) returns the floating-point number next to x in the direction of y. If $x = y$, `nextAfter` returns x.

The first two functions listed here are standard functions, while the more flexible `nextAfter` function is in math libraries for C and several other languages [16].

These operators will overflow to infinity, indicating overflow, and will underflow correctly as well. They also will only give one of the zeros, moving straight from -0.0 to the smallest subnormal number rather than going to 0.0. The rigorous definition is that `nextUp`(x) returns the smallest number that compares greater than x. The special cases for `nextUp` are enumerated in Table 8.1, with `nextDown`(x) behaving exactly like $-nextUp(-x)$. Aside from a signaling NaN input, `nextUp` and `nextDown` are silent.

The `nextAfter` function is less quiet in the C standard library, signaling an overflow exception when the function returns infinity and an underflow exception on a subnormal or zero output (in both cases with an inexact exception to boot). It also has behavior that `nextAfter`$(-0.0, +0.0) \Rightarrow +0.0$, returning y when x compares equal to y [16]. This behavior does not align with the standard `nextUp` and `nextDown`, but it comes from a different source. In C and the C-derived languages, `nextAfter` is the default iterator function, and the math library does not contain `nextUp` or `nextDown`.

Like many functions in this chapter, the iterators are actually implemented as short sequences of integer operations with some special case checks. In the

TABLE 8.1
Special cases of floating-point `nextUp`, mirrored in `nextDown`. Here, `TINY` represents the smallest positive subnormal number, while `HUGE` represents the largest positive non-infinite number.

Operation	Result	Exceptions
nextUp(± 0.0)	TINY	
nextUp($-$TINY)	-0.0	
nextUp(HUGE)	∞	
nextUp(∞)	∞	
nextUp($-\infty$)	$-$HUGE	
nextUp(NaN)	NaN	IO for sNaN

case of `nextAfter`, those special case checks raise exceptions. The core of `nextUp` and `nextDown` is either incrementing or decrementing a floating-point number as though it were an integer. Aside from the split between the negative and positive numbers, the mapping from integers to floating-point numbers is monotonic (recall Figure 2.5), so the easy way to get the next floating point number is to just get the next integer. This is shown in Algorithm 8.2, which describes a nearly branchless version of `nextUp` in 32-bit arithmetic.

Algorithm 8.2 Pseudocode for `nextUp` in Single-precision

Input: floating-point x
Output: floating-point $y = \text{nextUp}(x)$

1: **if** !isFinite(x) **then**
2:      **return** x
3: **end if**
4: $v \leftarrow x + 0.0$ $\triangleright$ Turn negative zero into positive zero
5: $i \leftarrow \text{bitCast}_{\text{I32}}(x)$
6: $s \leftarrow i \gg 31$ $\triangleright$ Get the floating-point sign bit from i
7: $d \leftarrow 1 - 2 \times (s)$ $\triangleright$ $d = 1$ if sign bit is 0, $d = -1$ if sign is 1
8: **return** $\text{bitCast}_{\text{F}}(i + d)$

These iterators are often used for testing floating-point functions when you would like to scan a part of the range of numbers. The main usage is to scan through a region to test the exact behavior of a piece of numerical code in that region. If you try the code for every possible input, you can guarantee that nothing unexpected will happen in that region. They can also tell you the uncertainty range of a given number by indicating the size of half a ULP in each direction:

$$u(x) = \frac{\text{nextUp}(x) - x}{2} + \frac{x - \text{nextDown}(x)}{2} = \frac{\text{nextUp}(x) - \text{nextDown}(x)}{2}$$

This function gives the size of the range of real numbers that would be represented by any floating-point number x.

The iterators are not generally used in runtime code, but for testing and verification, they can be useful tools for scoping out your precision and analyzing the cases you see. They also enable exhaustive testing of the floating-point number range if you happen to need it.

8.4 NaN Boxing

A significant amount of the encoding space of floating-point numbers is devoted to NaNs. There are $9\,007\,199\,254\,740\,990$ double-precision NaNs ($2^{53}-2$), just over nine quadrillion values. This is every value with exponent of all ones except for the ones with a zero mantissa. Looking at that encoding space, it is natural to think about what you might be able to do with it. There are two general classes of use for that encoding space: Usage to convey information about floating-point calculations—specifically when they go wrong—and usage to store things that are not floating-point. The former has been proposed by some math library developers, but is not in common use as of 2024. The latter usage is called **NaN boxing**, and is commonly used in heavily-optimized programs to make them more memory-efficient at a very small cost of arithmetic operations.

A key point about double-precision numbers is that there is enough encoding space in their NaNs to encode every possible pointer on a computer system. 2^{52} bits is enough encoding space to store byte addresses to 4 petabytes of memory, and at the time of writing, the largest computers on the planet have no more than 24 terabytes of RAM, and systems not built for extreme memory capacity stop at 1–2 terabytes. This means that as long as we can compress the 64-bit pointer values into a 52-byte encoding, we can store every floating point number *and* a pointer to every byte in an 8-byte slot. Pointers on 64-bit systems also are generally 8-byte aligned, so they will always have 3 trailing zero bits, leaving us encoding space for even more things. Increasing compression of memory like this comes at a cost of arithmetic operations and code that has more branches, but many programs are memory-bound and have arithmetic operations to burn.

This can be exploited by working with CPU and OS memory management. Most systems will give you all ones or all zeros in the top bits of a pointer, so compression is easy. Some systems will assign pointers with random 64-bit addresses or a tag in the top bits, and we need to be tricky if we are compressing pointers. The way around this is to use a customized allocator. Several programs that use NaN boxing for pointers will put those pointers inside a large region allocated with `mmap` or based on a single large initial allocation. This seems wasteful, but virtual memory will guarantee that this memory allocation does not actually occupy physical RAM until it is touched, so if you put a reasonable memory allocator on top of this big initial block of memory, you can use offsets within that block as your memory addresses, giving address compression. This is more difficult than using `malloc` or `new`, but it is a hack that allows you to significantly compress pointers by adding a small layer of indirection on your pointers.

A final benefit of NaN boxing is that you can do arithmetic on NaN-boxed values and anything that is not a number will give you an appropriate NaN. This allows arithmetic to be branchless. Many systems will propagate NaNs through calculations as well, completely preserving whatever it is you tried to operate on, meaning that you won't necessarily lose the payload. However, this is not universal outside the world of x86 CPUs, so operating non-destructively is better if you do this. Arm CPUs for example, don't preserve NaN payloads by default.

NaN boxing is most common with 64-bit values, but is not restricted to 64-bit values. there are $2^{24} - 2$ single-precision NaNs, which may not be enough to hold a pointer, but are enough to carry an object handle or an array index in addition to every floating-point value.

NaN Boxing in JavaScript Engines

The typical example of a system using NaN boxing to hold values is in browsers, where they use pointers stored in the NaN encoding space along with numbers stored in floating point. This allows quick identification of double-precision numbers and other types, and fast decoding. WebKit's JavaScript-Core (JSC) engine uses a variant of NaN boxing that stores pointers so that there is no modification to access pointers, but the encoding space of floating-point numbers is rotated. Firefox's JavaScript engine also uses NaN boxing, but a different form [6].

Normally, double-precision floating-point NaNs have an exponent of all ones, and quiet NaNs have an additional one bit. If we take the sign bit also as 1, that means we get NaNs when we have 13 leading one bits—the sign, the 11-bit exponent, and the quiet bit. Adding a 14th bit, we can now guarantee that we are only operating in NaNs with non-zero payloads, allowing a NaN with a zero payload in the encoding space.

JSC uses this space to store values with a 15-bit tag, and reserves tags of all ones *and all zeros* for things that are not floating-point numbers. In turn, the floating-point numbers are stored with an integer offset of 2^{49} from their canonical binary encodings, preserving every relevant floating-point number. The integer offset means that reserving a tag of zero for pointers takes NaN encoding space, not valuable encoding space near zero.

For JSC, a tag of zeros denotes a pointer or a special value. Values in binary are:

00000000 0000000P PPPPPPPP PPPPPPPP PPPPPPPP PPPPPPPP PPPPPPPP PPPPPPPP$_b$

This leaves 49 bits (the Ps) for pointers. Pointers themselves have tags in their three least significant bits, and special values including FALSE, TRUE, NULL, and UNDEFINED are also encoded in this space, using the second-least-significant bit to tag these special values. Zero is a banned encoding. The exact protocol used by this field is significantly more complicated around the tagging of the pointers, but pointers get preferred encoding for WebKit.

When the top 15 bits are not all ones or all zeros, you have a floating-point number that can be extracted by performing an integer subtraction of 2^{49} to re-align this range to start at zero. The encoding space for floating-point numbers before the subtraction is:

$$0002\ NNNN\ NNNN\ NNNN_h$$

$$\vdots$$

$$FFFC\ NNNN\ NNNN\ NNNN_h$$

In turn, after subtracting integer 2^{49}, this range of hexadecimal numbers corresponds to a ranges of double precision values that contain almost every useful value. These ranges are shown in Table 8.2. This leaves a space including all positive NaNs, both infinities, the zeros and subnormals, and all normal numbers. Every other use of the 8-byte value costs only a small slice of the negative NaNs.

Finally, a tag of all ones denotes a 32-bit integer. This is a speed optimization for the most common JavaScript numbers, which are 32-bit integers. Since JavaScript does a lot of integer math, but CPUs generally perform floating-

TABLE 8.2
Floating-point ranges available with WebKit's JavaScriptCore NaN boxing. These come after subtracting integer $2\ 0000\ 0000\ 0000_h$ (2^{49}) from the stored binary encoding.

Bit Fields of Floating-point Encoding			
Sign	Exponent (bin)	Mantissa (hex)	**Floating-point Value**
0	$111\ 1111\ 1111_b$	$F\ FFFF\ FFFF\ FFFF_h$	
	$\vdots$		All positive NaN
0	$111\ 1111\ 1111_b$	$0\ 0000\ 0000\ 0000_h$	∞
	$\vdots$		All positive numbers
0	$000\ 0000\ 0000_b$	$0\ 0000\ 0000\ 0000_h$	$+0.0$
1	$000\ 0000\ 0000_b$	$0\ 0000\ 0000\ 0000_h$	-0.0
	$\vdots$		All negative numbers
1	$111\ 1111\ 1111_b$	$0\ 0000\ 0000\ 0000_h$	$-\infty$
	$\vdots$		Some negative NaN
1	$111\ 1111\ 1111_b$	$9\ 0000\ 0000\ 0000_h$	

point calculations inside vector engines that make it difficult to use those values for tasks like indexing arrays, it is faster to separate out the integers when JSC can infer that numbers are being used as integers:

$$\mathtt{FFFE\,0000\,IIII\,IIII}_h$$

This allows the JavaScript engine to use 32-bit arithmetic instructions to do integer operations, ignoring the top 32 bits that have some junk in them.

This encoding allows pointers and integers to all be used unmodified, while all useful floating-point numbers are still accessible with a simple integer subtraction. The full specification for this encoding can be found in the WebKit repo on GitHub, including pointer tags and an encoding for small values stored in a BigInteger type. All of these fit inside eight bytes [22], allowing all of the normal numbers, fast integers, and machine pointers to occupy one value. This represents one of the most complex uses of NaN boxing, storing a tagged pointer, an integer, and all useful floating-point values in eight bytes. The rotation of the encoding space is a performance choice, as is the separation of integers. All of these choices add complexity but allow you to pack a lot of information into what otherwise would be wasted space.

Check Your Understanding

Problem 8.1. Write a function that performs $\mathtt{copyExp}(x, y)$, returning a floating-point number with the mantissa and sign of x, but the exponent of y.

Problem 8.2. Write a function that implements `nextAfter` using bit casting and integer operations. Make sure to handle the corner cases.
Challenge version: Make this function completely branchless.

Problem 8.3. Make a "double-or-string" type that can store either a double-precision floating-point number or a text string in one 8-byte value. Ensure that strings under 6 bytes long are stored as a small string within the 8-byte value, and that only larger strings are stored as a pointer.

Problem 8.4. Design a memory-efficient data structure for an ordered set of double-precision floating point numbers, using NaN boxing to store floating-point values for leaf nodes and pointers for internal nodes. Ensure that you have `add(x)`, `remove(x)`, and an iterator.
Challenge version: Use a B-tree (or B+ tree) with an arena allocator or a freelist for maximum speed.

9

Working with Error in Floating Point

Finally, we will take a look at several techniques to deal with floating-point error. Following a few simple rules can make accurate numerical calculations simpler to achieve. When those rules fail, it is possible to perform a basic error analysis on your calculation chains to find or rule out floating point issues. Another issue is the presence of non-numbers, like NaNs and infinities, that can creep in to calculations. Finally, if you don't care, it can help to enable fast math mode on a compiler or apply similar ideas for yourself, and we will discuss what "fast math" actually means.

9.1   Algebra for Accuracy

Some formulas compute equivalent things in the real numbers but are far more accurate in the floating-point numbers. A common example is the formula:

$$f(x) = \sqrt{x+1} - \sqrt{x}$$

As x grows, this formula causes a subtraction of two large numbers, which means cancellation. A plot of this function executed as written is shown as the left side of Figure 9.1.

A much more accurate formula that performs the same computation in the real numbers is given by using a difference-of-squares trick to rearrange things:

$$\sqrt{x+1} - \sqrt{x} = \frac{\left(\sqrt{x+1} - \sqrt{x}\right)\left(\sqrt{x+1} + \sqrt{x}\right)}{\sqrt{x+1} + \sqrt{x}} = \frac{1}{\sqrt{x+1} + \sqrt{x}}$$

The new formula here has no chance of cancellation, since we are adding two positive numbers and then dividing. It will be slightly slower due to the division, but for large x, is *far* more accurate. A plot of the new function is shown on the right side of Figure 9.1.

These sorts of algebra tricks to remove subtractions and additions that can have opposite signs are good ways to make calculations more precise. The quadratic formula is another famous example of a formula that does not work

$$f(x) = \sqrt{x+1} - \sqrt{x} \qquad\qquad f(x) = \frac{1}{\sqrt{x+1}+\sqrt{x}}$$

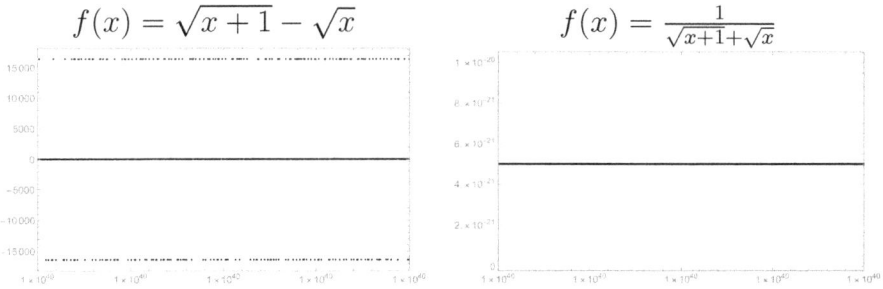

FIGURE 9.1
Plots of two functions that compute $\sqrt{x+1} - \sqrt{x}$ in double precision. Left: Plot the naive function, which jitters randomly around zero, giving spurious results of $\pm 16\,384$. Right: Plot of a corrected function that gives the correct result of 5×10^{-21} across this window.

as well in floating point as you might hope:

$$\frac{-b \pm \sqrt{b^2 - 4ac}}{2a}$$

This formula has two additions that can have opposite signs, one of which is guaranteed to have an opposite sign since it is a "$\pm$" that represents an addition and a subtraction. The $b^2 - 4ac$ inside the square root can also cause cancellation when a and c have the same sign. Some algebra will produce a significantly better formula depending on the values of a, b, and c, but the best formula to use depends on the magnitudes and regimes you are working with. Any chance you have to turn possible subtraction into same-sign addition and/or any multiplicative operation will likely improve accuracy.

Tools like **herbie** can be used to do some of this algebraic manipulation automatically [23]. Herbie analyzes floating-point expressions for accuracy across their range and determines more accurate substitutions that are algebraically equivalent in the real number space. However, they are limited to a fixed set of algebraic manipulations and cannot change the inputs or outputs of the formulas used. Put another way, Herbie is probably better at coming up with the correct algebraic tricks than we are, but only views the narrow environment of a single floating-point expression.

A final class of algebraic manipulations that can help with accuracy is to perform changes of coordinates. When you need to do a lot of manipulations on a quantity that is large or small, it can help to center on that quantity by doing a linear change of coordinates for that calculation. An example of a two-dimensional grid represented in a floating-point format is given in Figure 9.2, showing the effects of being close to or far from the origin.

An implied change of coordinates that is common to see in simulations is calculation of a delta to apply to an existing number instead of directly

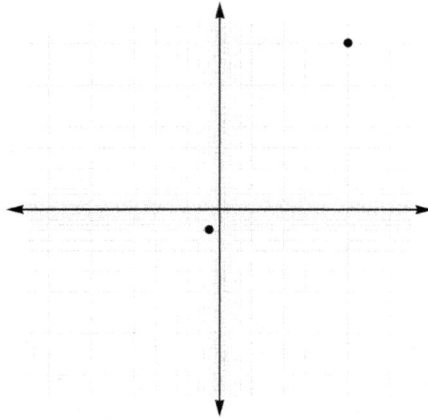

FIGURE 9.2
Grid showing the resolution of two-dimensional space in floating point. Around the point in the upper right side of the grid, the resolution of the grid is significantly lower than it is at the point near the origin.

calculating a new number. Operations update a change in a quantity, which often defaults to a smaller number than the quantity itself, only updating the original number after computing the final delta. Errors of several ULPs in a delta that is close to zero become an exact operation when you finally add that delta to a quantity in the thousands. This kind of transformation can also allow you to take advantage of numerical integration methods (see Chapter 12) in cases where you are updating a number over time, since you are creating a differential equation by separating out the "velocity" of that number into a delta.

Changes of coordinates are routinely used in open-world and massively-multiplayer games, where the technique is called **floating origin**. Floating origin will center the coordinate $(0, 0, 0)$ on the camera or player, allowing fine precision for local movement. This also has the effect of keeping the math and associated simulations consistent no matter where the player is located in the in-game world. Floating origin also allows the player's view to be rendered using reduced precision while the state of the world is stored at higher precision with no loss of accuracy (e.g., a world whose state is stored in double-precision on the server side can be rendered quickly and accurately in single-precision).

Additionally, some problems benefit from alternative coordinate systems. Problems that are circular in nature, such as planetary orbits, can use polar coordinates to have a nice closed-form solution rather than running a full simulation. Kepler's equations of motion are simple to define in polar coordinates, while expressing them in terms of points on an x-y grid is comparatively messy. Translating from polar coordinates to usable Cartesian coordinates and back takes a few trigonometric operations. Similarly, simulations with rotational symmetry of some kind are often much more simply expressed in polar coordinates than on a grid.

Averaging Floating-point Sets

Computing averages of large datasets is more challenging than it seems since you are essentially computing a quantity that looks like a very large sum. When averaging a large sample of numbers that are about the same size, the typical formula for computing an average is:

$$E = \frac{1}{N} \sum_{n=1}^{n} x_n$$

The running sum of a large set is hard to compute accurately and risks overflow. There are algorithms, such as a famous one that uses double-word arithmetic (we will revisit this example in Section 9.3) for taking accurate sums, but this is often not necessary unless N gets very large.

We can reformulate our calculation into an iterative calculation that keeps a running estimate of the average, E_n, rather than keeping a sum:

$$E_{n+1} = \frac{n}{n+1} E_n + \frac{x_n}{n+1}$$

At each step, we adjust the current average by a small amount and then add a small adjustment corresponding to x_n. We still have problems as n gets large, since the $\frac{x_n}{n+1}$ term gets smaller and smaller and $\frac{n}{n+1}$ loses precision as it approaches one, but this formula now will not overflow for large x values. However, in floating point we have rounding, which means that:

$$\frac{n}{n+1} + \frac{1}{n+1} \neq 1$$

Our old formula actually gave each round a weight that is slightly more or less than one, which can cause systematic error depending on N. A new formula that does guarantee a total weight of one in each round is:

$$F_n = \frac{n}{n+1}$$
$$E_{n+1} = F_n E_n + (1 - F_n) x_n$$

Our subtraction here *will* always be exact because F_n is always close to one (for the theoretically inclined, this is shown by Sterbenz's lemma [24]). If we had instead computed F_n as the weight for x_n and used $1 - F_n$ as the coefficient of E_n, we would not have been able to say that this subtraction is always exact. Practically, this formula will produce inaccuracy by jittering the weight of each x a bit depending on whether the F_n calculation rounded up or down. When n is small enough, this jitter is essentially random and will often average out. As F_n gets closer and closer to one, however, this jitter becomes biased and the usability of this formula breaks down.

A popular formula for the average of a set of numbers that avoids this problem with large n is:

$$E_{n+1} = E_n + \frac{x_n - E_n}{n + 1}$$

By now calculating using a delta adjustment, we avoid the error we would have gotten from $\frac{n}{n+1}$. However, we have added a subtraction to a place where we previously did not have one, creating a new source of precision loss due to cancellation, and we are now performing an addition with a larger disparity in order of magnitude than we had before. Previously, we were adding a quantity close to $\frac{E}{n+1}$ to a number close to E, but now we are adding $\frac{x-E}{n+1}$, which is a significantly smaller quantity.

It is always possible to also parallelize the calculation of means, too, but we should parallelize with a numerically stable formula rather than parallelizing to compensate for the use of a bad one. Additionally, lists sorted by magnitude will look somewhat better than lists in random order, but there are more efficient ways to get a clean average of a set when this is not the case, falling back to the naive algorithm for averaging, but using clever tricks to produce an accurate sum.

9.2 Changes of Units

As we have seen from the Minecraft example and the example of trigonometric functions, changes of units alone can be a way to avoid accumulation of error in floating-point calculations. Whenever inexact constants are repeatedly used, error from those constants has a tendency to accumulate. It is far better to choose units in which common values and constants can be represented exactly. The most classic example of a bad unit is using floating-point dollars for financial calculations. Only four out of 100 possible values between $1.00 and $1.99 can be represented exactly in floating point. A far better unit for financial calculations is to work in cents or in hundredths of cents. The same logic goes for angles; using units of half-revolutions will make common angles (like right angles) exact, while they are approximate in radians.

In video games and simulators, it is often an option to choose to use weirder units rather than physical quantities. The player's perspective of a distance of one unit can vary a lot from game to game, from being a small piece of a character to the size of a full room. However, representing things like the player velocities and gravitational constants in terms of "units per frame" or "units per frame²" can cause small errors between the intent of the designer and the implementation of the game. The use of units that allow exact representation of common quantities allows you to start off from a baseline that is as numerically clean as possible, and allow error to accumulate only through rounding rather than contending with systematic errors.

Unit-related bugs don't just occur in games like Minecraft. The United States' Patriot missile system uses a 24-bit representation of time, counted in units of 0.1 seconds. When calculating the trajectories of incoming missiles to intercept, the Patriot's software used floating-point seconds as its unit of time, calculated by multiplying its integer time counter by a single-precision representation of 0.1. As we know, 0.1 cannot be exactly represented in floating point, so the value they used was rounded down, causing all times in the Patriot system to have a systematic error of approximately 0.0001%. This was not normally a problem since these machines were frequently shut off and redeployed, but as error accumulates over time, measurements of short durations in the Patriot system become very inaccurate. This, in turn, caused Patriot systems that had been deployed for a long time to lose track of fast-moving targets. In February of 1991, a Patriot system that had been running for 100 hours lost track of a Scud missile due to this time error and failed to intercept it, causing 28 soldiers to lose their lives [25]. Working in units of "meters per tenth of a second" instead of meters per second would have avoided this issue.

In a different vein, the units used can also affect when overflow and underflow happen. Half of all floating-point numbers have a magnitude less than one, and half have a magnitude greater than one. If you have a system with a very wide dynamic range, it can be more useful to set your units so that numbers are usually around one, allowing the numbers to grow and shrink as much as possible while avoiding underflow or overflow.

Being careful about units in general is a good idea in any form of numerical software. Another famous numerical spacecraft accident, the Mars Climate Orbiter crash, was related to bad use of units. The spacecraft was intended to use metric units, but one contractor who supplied software for the spacecraft used US customary units instead. With this unit discrepancy, the spacecraft quickly fell out of its orbit and crashed [26]. This is not a numerics bug in the traditional sense, but it underscores the importance of making sure your units make sense and are used consistently through the entire system.

The need to track units adds programmer overhead, and this creates a source of errors even when your math is correct and you use a memory-safe programming language. Programming languages and library developers have added add libraries that allow you to encode physical units as language types so that unforced errors involving units will be harder to introduce [27].

9.3   Double-word Arithmetic

A conceptually simple way to get more precision is to widen the numeric format you are using, but when you run out of fast numeric formats that are available, double-word arithmetic offers an attractive speed-accuracy tradeoff. Double-word arithmetic increases your precision by using two floating-point numbers, a coarse number and a fine number, to represent one quantity. The full value of the number is equal to the sum of the coarse and fine parts. This allows you to get precision close to quad precision while still doing math on double-precision numbers or precision close to double while using single-precision arithmetic. A colloquial term to refer to double-word arithmetic with double-precision operands is **double-double** arithmetic, referring to the fact that two doubles are used for each quantity.

Double-double arithmetic offers 106 bits of precision, ($p = 53$ for double precision, so 53×2), which is almost as precise as a quad-precision number. However, the dynamic range does not change, and a significant chunk of the number space has reduced precision. The exponent of the coarse value in the pair carries the exponent of the number, so double-double numbers cannot have larger magnitude than doubles. The exponent of the fine value is at most p less than the exponent of the coarse value, so the minimum exponent where double-double numbers have full precision is p higher than it is for doubles. Below that, the fine value becomes a subnormal number, and the precision of double-doubles degrades. However, there is no obligation that the fine part of the number has exactly an exponent that is p less than the coarse exponent, so it is possible to represent quantities like $1 + 2^{-200}$ exactly in double-double arithmetic.

There are a few libraries available that perform double-double arithmetic, and can go up to quad-double arithmetic, offering 212 bits of precision. The QD library provides basic operations and some transcendental functions in double-double and quad-double arithmetic in C or C++ [28], and is the most commonly used library for this form of arithmetic. These libraries back most uses of double-word arithmetic.

Double-word arithmetic is built on top of a set of three operations that create perfect two-output sums and products of floating-point numbers. These operations are collectively referred to as **augmented operations**. There are two operations, fast2sum and 2sum, which are used to perform augmented sums of floating-point numbers. The fast version is used when it is known which operand is larger, while the slower version can handle operands of unknown order. The third operation performs an augmented product of two floating-point numbers, and is called 2prod.

Algorithm 9.1 shows the fast two-output sum, which is done by adding the numbers and then finding the error of that addition to put in the low part. All of these operations are built out of round-to-nearest floating-point operations, which we make explicit in the following algorithms. Double-word arithmetic algorithms are all about handling the error associated with rounding, so it is worth a note to be explicit about exactly where rounding occurs.

Algorithm 9.1 Fast Two-output (Augmented) Sum "`fast2sum`"

Input: Double-precision a, b with $|a| \geq |b|$
Output: Double-double $x = (x_h, x_l)$ equal to $a + b$

1: $x_h \leftarrow \text{RTN}(a + b)$
2: $t \leftarrow \text{RTN}(x_h - a)$ ▷ Find the error in the sum we just took
3: $x_l \leftarrow \text{RTN}(b - t)$
4: **return** (x_h, x_l)

The generic version of the `2sum` algorithm, described in Algorithm 9.2, does a few extra floating-point operations in order to avoid a branch. It is possible to check for the larger number and then swap if needed, but that is usually slower on processors than adding more arithmetic operations. The computation starts the same as the fast algorithm, but we adjust the low term to handle both $|a| > |b|$ and $|a| < |b|$.

Algorithm 9.2 Generic Two-output (Augmented) Sum "2sum"

Input: Double-precision a, b
Output: Double-double $x = (x_h, x_l)$ equal to $a + b$

1: $x_h \leftarrow \text{RTN}(a + b)$
2: $t_g \leftarrow \text{RTN}(x_h - a)$ ▷ Find the error in the sum if $a > b$
3: $t_l \leftarrow \text{RTN}(x_h - t_g)$ ▷ Start delta error if $b > a$
4: $g \leftarrow \text{RTN}(b - t_g)$ ▷ Finish error if $a > b$
5: $l \leftarrow \text{RTN}(a - t_l)$ ▷ Finish error if $b > a$
6: $x_l \leftarrow \text{RTN}(g + l)$ ▷ Final construction of x_l
7: **return** (x_h, x_l)

The extended product is trivial with the FMA operation, first doing a coarse product, then subtracting that coarse product from $a \times b$ to get the error. The resulting algorithm is shown in Algorithm 9.3. The FMA operation also rounds to nearest in this algorithm.

Algorithm 9.3 Two-output (Augmented) Product "2prod"

Input: Double-precision a, b
Output: Double-double $x = (x_h, x_l)$ equal to $a \times b$

1: $x_h \leftarrow \text{RTN}(a \times b)$ ▷ Get the coarse product
2: $x_l \leftarrow \text{FMA}(a, b, -x_h)$ ▷ Use FMA to get the tail as $a \times b - x_h$
3: **return** (x_h, x_l)

With these basic augmented operations, we can then start to construct double-word operations. The recent version of the IEEE 754 specification includes versions of these operations that give better accuracy and have slightly different rounding, but at the time of writing, no hardware vendor implements these operations. It might therefore be more precise to not use the term "augmented" to describe the 2sum and 2prod algorithms, but they fill the same roles that these standardized augmented operations eventually will.

Building on top of our basic operations, the algorithm to take the product of two double-word numbers is shown Algorithm 9.4, and the sum of double-word numbers is shown in Algorithm 9.5. Both of these can be significantly simplified when one operand is a singular double-precision number rather than a double-double [29]. The general schema of both algorithms is to compute the partial results that must be added together to produce the final result, and then use the fast2sum algorithm to add each of these parts into the final result. Both algorithms presented here are optimized, leaving out terms that do not have a significant effect on the final output.

Algorithm 9.4 Double-double Product

Input: Double-double $a = (a_h, a_l)$, $b = (b_h, b_l)$
Output: Double-double result $p = (p_h, p_l)$ equal to $a \times b$

1: $(c_h, c_l) \leftarrow \mathtt{2prod}(a_h, b_h)$ ▷ Full high-high product for p_h
2: $l_{ab} \leftarrow \mathtt{RTN}(a_h \times b_l)$ ▷ High-low products for p_l
3: $l_{ba} \leftarrow \mathtt{RTN}(a_l \times b_h)$
4: $x \leftarrow \mathtt{RTN}(l_{ab} + l_{ba})$ ▷ Add up terms for p_l into y
5: $y \leftarrow \mathtt{RTN}(x + c_l)$
6: $(p_h, p_l) \leftarrow \mathtt{fast2sum}(c_h, y)$ ▷ Add in the y error term
7: **return** (p_h, p_l)

Algorithm 9.5 Double-double Sum

Input: Double-double $a = (a_h, a_l)$, $b = (b_h, b_l)$
Output: Double-double result $s = (s_h, s_l)$ equal to $a + b$

1: $(h_h, h_l) \leftarrow \mathtt{2sum}(a_h, b_h)$ ▷ High-high sum
2: $(l_h, l_l) \leftarrow \mathtt{2sum}(a_l, b_l)$ ▷ Low-low sum
3: $c \leftarrow \mathtt{RTN}(h_l + l_h)$ ▷ Carry from high part of low sum to the result
4: $(t_h, t_l) \leftarrow \mathtt{fast2sum}(h_h, c)$
5: $d \leftarrow \mathtt{RTN}(l_l + t_l)$ ▷ Carry from low part of low sum to the result
6: $(s_h, s_l) \leftarrow \mathtt{fast2sum}(t_h, d)$
7: **return** (s_h, s_l)

While these are the basic algorithms for sums and products, other algorithms, including division and transcendental functions, can be constructed relatively efficiently in double-word arithmetic when compared to software emulation of a wider floating-point format. When you absolutely need more precision, they are good default options. Note that the product of two double-doubles only needs nine floating-point operations (with FMA being one), while double-double sum needs 20 floating-point operations. All of these are branch-free, so they "look" to a processor like normal arithmetic. They also handle overflow and underflow correctly, but with the normal caveats on the reduced dynamic range of double-double compared to double precision.

Double-word operations are traditionally used only when there are no hardware operators otherwise available, but simulations on gaming GPUs may be able to make use of them as well. Many gaming GPUs leave you with significantly reduced double-precision calculation capabilities in comparison to their single-precision capabilities, with recent GPUs offering 32 or 64 times fewer double-precision FLOPS than single-precision FLOPS. If you happen to need precision close to double precision but cannot pay the cost of double precision on your GPU, using double-word arithmetic on single-precision numbers may be a viable alternative, giving 48 bits of precision with each primitive taking less than 32 single-precision floating-point operations.

Double-word algorithms all rely on the precise order of calculations, so the use of double-word arithmetic limits your options for optimizing other math. In particular, some compiler options (see Section 9.7) can cause these algorithms to be completely optimized away since they may not preserve exact ordering. In the course of the `fast2sum` operation, we perform the following floating point operations to get x_l:

$$b - ((a + b) - a) \Rightarrow x_l$$

If our compiler thinks that floating-point math is associative, this expression completely reduces to $0 \Rightarrow x_l$, which completely invalidates most of double-word arithmetic.

Double-word Arithmetic in Sums of Sets

When we discussed calculation of averages earlier in the chapter, the simple algorithm was to take a sum of the set of numbers. However, those sums risk losing precision. If we add some extra precision to our running total as we take the sum, we can avoid some of this loss by simply having more precision that is available to lose.

An algorithm that makes use of a limited form of double-word arithmetic to add extra precision to sums is Kahan's summation algorithm, also known as compensated summation. This is an algorithm for taking an accurate sum of double-precision numbers using double-word arithmetic, although it was introduced several decades before double-word arithmetic was a known technique. This algorithm is not perfect, but it stands up to significantly larger numbers of elements than a simpler running sum due to the increased precision of the accumulator. The algorithm is given in Algorithm 9.6 [30].

Algorithm 9.6 Compensated Summation

Input: List of doubles, $X = \{x_0, x_1, ... x_N\}$
Output: Double-precision s equal to the sum of all elements of X

1: $s \leftarrow 0.0$ ▷ Initialize
2: $c \leftarrow 0.0$
3: **for** $x_i \in X$ **do**
4:      $y \leftarrow x_i + c$ ▷ Add our compensation term to x_i
5:      $(s, c) \leftarrow$ `fast2sum`(s, y) ▷ Sum and get new compensation term
6: **end for**
7: **return** s

We keep the running rounding error of summation in c, and each time we go to add a new number, we add c to that number before doing a double-double addition of that number and our final sum. Note that if our final sum could be smaller in magnitude than y while requiring full precision, we may need to replace `fast2sum` with the more generic `2sum` in this algorithm. This is what can happen if we get significant cancellation, in which case the use of `fast2sum` in this algorithm can add some error compared to `2sum`. In almost every case, though, $|s| > |c|$ will hold.

This algorithm doubles the precision of the accumulator when we take the sum of a list of floating-point numbers, but due to the wide dynamic range of floating point, that is not enough to get a fully accurate and reproducible sum. This algorithm, like all summation algorithms, can also overflow readily since the dynamic range of our double-double accumulator is no different than the dynamic range of x. This algorithm does mostly prevent error from cancellation during the summation process, though, since the compensation field will carry the tail bits if we do hit an instance of a large positive sum adding a similarly large negative x_i. Doing much better than this requires a completely different approach to taking a sum.

The ultimate solution to the problem of summation of gigantic lists of floating-point numbers involves binning the numbers in the set by magnitude, and summing using accumulators in each bin, jumping up to the next bin any time a bin's sum gets too big. The most extreme version of this is to use one accumulator for each exponent value, tracking over 2 000 separate accumulators for a double-precision sum. Compromise versions with hundreds of accumulators spread over tens of bins work almost as well (the mapping of bin to accumulator need not be one-to-one). At a cost of a significant amount of complexity, memory, and speed, you can approach a correct sum of a set of numbers. This type of algorithm is so expensive that it is rarely used except when exact reproducibility of floating-point calculations is desired [31]. Unless your set is pathological, double-word arithmetic will deliver a correct sum almost all the time.

9.4 Interval Arithmetic

When you need to know exactly where a real-numbered quantity could be in a floating-point calculation chain, interval arithmetic is the solution. Interval arithmetic is precisely vague: it will not tell you exactly where a number is, but it will tell you exactly where that number is *not*.

Like double-word arithmetic, interval arithmetic uses two floating-point numbers to represent each quantity, but the meaning of these numbers is different. For interval arithmetic, the pair of numbers instead represents the minimum and maximum of a confidence interval. Operations in interval arithmetic are performed so that the interval continues to represent the outer limits of where that quantity might be. The lower end of the interval and the upper end of the interval are moved independently based on the operation in question. Intervals are generally constructed as closed intervals, with the exception of zero which gets special treatment: An interval containing both zeros is a closed interval at zero, while an interval containing only one of the zeros is an open interval at zero.

For example, the sum of two numbers in interval arithmetic is the interval between the sum of the lower ends of the interval and the sum of the upper ends of the interval:

$$[2, 3] + [1, 4] \Rightarrow [3, 7]$$

Products of intervals can behave similarly, although negative numbers in intervals can cause this to need a bit of care:

$$[-3, 4] \times [1, 7] \Rightarrow [-21, 28]$$

The ceiling of the interval becomes the maximum of the products of endpoints of the respective intervals, while the floor of the interval is the minimum of the products of endpoints. Operations on pairs of intervals always produce a wider result than the inputs, while operations that involve a constant (or some unary operations) can make the interval narrower:

$$1 + 0.02 \times [1, 2] \Rightarrow [1.02, 1.04]$$

Rounding modes for interval arithmetic also need to be set to handle the "precisely vague" nature of interval arithmetic. The lower endpoint of the resulting interval must be rounded down, and the upper endpoint of the resulting interval must be rounded up:

$$[0.01, 0.03] \times [-0.4, 3.0] \Rightarrow [\text{RD}(0.03 \times -0.4), \text{RU}(0.03 \times 3.0)]$$

This use of rounding modes keeps the contract that the real-valued result must be within that interval, but there is a general tendency for intervals in interval arithmetic to widen over a calculation chain, and rounding is one of the fundamental mechanisms that makes that happen.

Due to the need to switch rounding modes and the corner cases around operations, interval arithmetic is slow. It is slower than double-word arithmetic, which does not require rounding mode switching, and many operations contain branching conditions that can cause CPUs to slow down significantly. However, interval arithmetic can be helpful in testing and development environments, where it can help to identify points where computations introduce uncertainty into the final results of a calculation. Herbie, the numerical accuracy tool from Section 9.1, uses interval arithmetic internally for this purpose.

Interval arithmetic can also identify the ranges of calculations, allowing you to do further optimization. When given a limited input space, many calculations have a significantly restricted output space. This can work to your advantage to simplify expressions, use narrower formats, or create approximations targeted to specific regions. Similarly, if you find that your output space is wider than you expect, interval arithmetic helps you identify where that happens and where you might need to insert a check for an infinity or a NaN.

9.5   Error Analysis

Error analysis is the bread and butter of many parts of numerical analysis, and it is an available tool for developers if needed, but is relatively mathematically trick. There are two types of error analysis that answer the question "how bad could my calculation possibly get?"

With $y = f(x)$ as computed in floating point, forward error analysis looks for Δy such that:

$$\Delta y = |y - f(x)|$$

In other words, a forward error analysis looks at the error that could show up in the calculation of y compared to the ideal $f(x)$. In practice, this type of analysis can be difficult to perform, and may also be somewhat meaningless since you have to track the precision of $f(x)$ itself to know if your forward error is within acceptable bounds.

A simpler (in floating point) and more useful measure is given by a backward error analysis, which looks for Δx such that:

$$f(x \pm \Delta x) = y$$

In other words, we look at the implied error in the input that is given by calculating y. We have seen the concepts of rounding and ULPs of precision, so we perform a backward error analysis by backtracking those errors through the calculation of $f(x)$. This also allows us to compare Δx to the precision we know we have for x.

We can perform a backward error analysis by treating rounding as though it adds some error, ϵ, to the calculation. That rounding error is equal to half of a ULP for the basic operations. The transcendental functions from Chapter 7 will have ϵ determined by the accuracy of the math library used (see Chapter 10). So, when f is a simple product, we get Δx for each input by:

$$f(a, b) = (1 + \epsilon)(a \times b) = [a(1 + \epsilon)] \, b = a \, [b(1 + \epsilon)]$$

Thus, the backwards error analysis of this product implies $\Delta a = \epsilon a$ and $\Delta b = \epsilon b$. Since ϵ is relative and equal to half a ULP, the implied error on a is half a ULP. For a more complicated example, consider the difference of square roots from Section 9.1:

$$f(x) = \sqrt{x + 1} - \sqrt{x}$$

Adding an epsilon to propagate backward at each point of rounding:

$$y \approx (1 + \epsilon_1) \left[(1 + \epsilon_2)\sqrt{(1 + \epsilon_3)(x + 1)} - (1 + \epsilon_4)\sqrt{x} \right]$$

Pushing down toward x:

$$y \approx (1 + \epsilon_1)(1 + \epsilon_2)\sqrt{(1 + \epsilon_3)(x + 1)} - (1 + \epsilon_1)(1 + \epsilon_4)\sqrt{x}$$
$$= (1 + \epsilon_1 + \epsilon_2 + \epsilon_1\epsilon_2)\sqrt{(1 + \epsilon_3)(x + 1)} - (1 + \epsilon_1 + \epsilon_4 + \epsilon_1\epsilon_4)\sqrt{x}$$

Since ϵ values are small, we will drop the second-order terms $\epsilon_1\epsilon_2$ and $\epsilon_1\epsilon_4$, as well as dropping subsequent second-order terms:

$$y \approx (1 + \epsilon_1 + \epsilon_2)\sqrt{(1 + \epsilon_3)(x + 1)} - (1 + \epsilon_1 + \epsilon_4)\sqrt{x}$$

If we want to think about error when x is large, we can use the simplifying assumption that $(x + 1) \approx x$:

$$y \approx (1 + \epsilon_1 + \epsilon_2)\sqrt{(1 + \epsilon_3)x} - (1 + \epsilon_1 + \epsilon_4)\sqrt{x}$$
$$= \left[(1 + \epsilon_1 + \epsilon_2)\sqrt{(1 + \epsilon_3)} - (1 + \epsilon_1 + \epsilon_4) \right]\sqrt{x}$$
$$= \sqrt{\left[(1 + \epsilon_1 + \epsilon_2)^2(1 + \epsilon_3) + (1 + \epsilon_1 + \epsilon_4)^2 - 2(1 + 2\epsilon_1 + \epsilon_2 + \epsilon_4)\sqrt{1 + \epsilon_3} \right] x}$$

Simplifying the ϵ terms while dropping all high-order terms:

$$y \approx \sqrt{\left(2 + 4\epsilon_1 + 2\epsilon_2 + \epsilon_3 + 2\epsilon_4 - 2\sqrt{1 + \epsilon_3} \right) x}$$

Since ϵ_3 is small, we can even get rid of the radical, as $\sqrt{1 + \xi} \to 1$ for small ξ:

$$y \approx \sqrt{(4\epsilon_1 + 2\epsilon_2 + \epsilon_3 + 2\epsilon_4)\, x}$$

We can now see the problem. There is no factor of x left that determines y! Rethinking this in terms of $f(x + \Delta x)$, we can see that $\Delta x \approx -x$. To first order, x doesn't matter in terms of producing a result. This mathematically confirms what we saw in Figure 9.1.

After we do the algebra from Section 9.1, noting that $f(x)$ in real terms can also be written as $f(x) = \frac{1}{\sqrt{x+1}+\sqrt{x}}$, our error analysis looks somewhat different:

$$y = \frac{1+\epsilon_1}{(1+\epsilon_2)\left[(1+\epsilon_3)\sqrt{(1+\epsilon_4)(x+1)}+(1+\epsilon_5)\sqrt{x}\right]}$$

$$\approx \frac{1+\epsilon_1}{(1+\epsilon_2)\left[(1+\epsilon_3)\sqrt{1+\epsilon_4}+1+\epsilon_5\right]\sqrt{x}}$$

$$\approx \frac{1+\epsilon_1}{(1+\epsilon_2)(2+\epsilon_3+\epsilon_5)\sqrt{x}}$$

$$\approx \frac{1+\epsilon_1}{(2+2\epsilon_2+\epsilon_3+\epsilon_5)\sqrt{x}}$$

$$= \frac{1}{\sqrt{\frac{(2+2\epsilon_2+\epsilon_3+\epsilon_5)^2}{(1+\epsilon_1)^2}x}}$$

We get the expression in the radical to find Δx:

$$\frac{(2+2\epsilon_2+\epsilon_3+\epsilon_5)^2}{(1+\epsilon_1)^2}x \approx \frac{4+4\epsilon_2+2\epsilon_3+2\epsilon_5}{1+2\epsilon_1}x$$

This expression is approximating $4x$, since $\sqrt{x+1}+\sqrt{x} \approx 2\sqrt{x}$ under our large x assumption.

Expanding, we will find that the difference between 4 and the expression of ϵ here is a function of ϵ and is not large, and we can also note that none of the ϵ values here will be small. All five of the operations here will give half a ULP of error, since the addition has two values of the same order of magnitude and all of the other expressions are multiplicative or exact unary operations (square roots). The only epsilon value that threatens to be large is ϵ_4, which is a second-order piece of error either way. For double precision, $\epsilon = \pm 2^{-53}$, so our outer bound on x, which gives Δx, comes from dividing the previous expression by four and plugging in the worst-case values of ϵ:

$$\Delta x \approx \left|1 - \frac{4+3\times2^{-52}}{4(1-2^{-52})}\right|x = (4\times10^{-16})x \approx 2 \text{ ULP}$$

This gives our implied error in x from performing this calculation. In a sense, we are adding 2 ULPs of error to our input by performing this calculation or removing two bits from the precision of the input.

Error analysis of this form is essentially a manual application of a symbolic form of interval arithmetic. The algebra can get tricky, but it is possible, using simplifying assumptions, to get a reasonable first-order idea of what your error would be from a set of calculations.

9.6   Avoiding Invalid Operations

When working with floating point, it is natural to think about small errors, but accidental invalid operations are one of the more spectacular and visible ways that floating-point calculations can return a result that you would prefer not to see. In many ways, returning a NaN when you shouldn't is the ultimate numerical inaccuracy.

Getting a NaN out of a calculation that shouldn't produce one is usually a two-step process:

1. Do a numerical operation that yields either a zero or an infinity (or in some cases, a negative number).

2. Pass that result into a calculation that can't accept it.

Looking through the tables of special cases in Chapters 4–7 and referring back to Table 3.1, it is easy to see how this pattern arises. The first step here is often the result of an overflow or underflow, although it can be a valid result. The next step is what turns that result into a problem. Once you have that NaN, though, it will propagate through the rest of your calculations with very few exceptions. Specifically, of the operations mentioned in the previous chapters, only the **pow** family of functions and the **minimum/maximumNumber** functions can swallow a NaN.

Many of the techniques discussed in this chapter can help avoid invalid operations, but it is often up to you to figure out where these could occur. Interval arithmetic can play a role here, showing you exactly when an interval can produce an infinity or an otherwise unexpected value, if you know the corners of the input of your calculation. This can reveal things like $asin(x)$ returning a value greater than π (remember that irrational numbers get rounded) or other similar arithmetic issues. However, many cases of this sort of issue show up when numbers exceed the expected range.

Speed and safety have a natural tension in software. Checking bounds at each numerical operation is often prohibitively expensive. However, a natural place to put your bounds checks is at the borders of calculations. Checking that the inputs are in an expected range or checking for problems at the outputs allow you to centralize handling. NaN propagation is not necessarily a problem within a floating-point calculation, but becomes a problem any time you try to do anything with the result.

An fun recent example of this sort of overflow comes to us from the video game *Balatro*. *Balatro* uses double-precision numbers for its scores, but it has a scoring system that is replete with multipliers such that the user can, with some difficulty, even overflow the double-precision score. The game displays animations whose size and position is relative to the player's score, and these animations glitch and overflow past their boundaries in the user interface when player's the score overflows to infinity.

9.7   Fast Math Compiler Options

When you do not need precision, compilers have a number of optimizations that they can do that cause error but make your math faster. What is included in these fast math settings depends on a balance of speed and accuracy, so they can be nebulously defined. An "acceptable" optimization in one compiler may be unacceptable in another compiler or language. Note that these optimization options are usually disabled unless you request them, even when you set the compiler to heavily optimize your code, because they change the semantics of your code.

Many compilers follow the GNU C compiler in terms of what is an acceptable transformation for fast math. The settings that the GCC turns on are the following [32]:

- `-fno-math-errno` disables reporting of floating-point exceptions in the C `errno` field. This is specific to C and C++.

- `-ffinite-math-only` causes the compiler to assume that numbers are always finite, and neither infinity nor `NaN`. This removes a lot of special-case checks in math libraries.

- `-fno-rounding-math` allows the compiler to assume that rounding modes will not change throughout the program, allowing it to do things like constant folding.

- `-fno-signaling-nans` disables handling of `sNaN`.

- `-fexcess-precision=fast` allows the compiler to move your calculations to a wider floating-point format if that will make them faster. The added spurious widening and narrowing of the number format can cause extra rounding.

- `-fcx-limited-range` simplifies division of complex numbers.

- `-funsafe-math-optimizations` enables several optimizations that are both fun and safe. This flag is a combination of several unsafe options in itself:

    - `-fno-signed-zeros` disables the use and special handling of negative zero. You can still sometimes get negative zeros in this mode, but operators treat them identically to positive zero.

    - `-fno-trapping-math` disables trapping on floating-point exceptions (even the bad ones).

    - `-fassociative-math` causes the compiler to assume that floating-point math is associative, enabling the compiler to do algebra as though the numbers in question were real numbers (i.e., everything discussed in Section 4.6 is now in the compiler's hands).

— `-freciprocal-math` allows the compiler to compute and use the reciprocal of a value instead of using division.

Some of these flags are usually appropriate in numerical code, but not always all of them. The flags that specify error handling behavior are often usable in optimized code, where you might prefer to ignore floating-point errors to begin with. Most programs do not change their rounding mode from the defaults, so `-fno-rounding-math` can usually be safe. Finally, signaling NaNs are designed to be injected by software and won't be produced by floating-point functions, so if you do not plan to inject any, `-fno-signaling-nans` can be a small performance boost at no cost. Many of the remaining flags in this list either explicitly trade off accuracy, such as the flags for associative math and reciprocals, or restrict the range of usable floating-point values to make computations faster.

At the time of writing, the Clang compiler goes further than GCC in terms of fast math options, also adding an option (`-fapprox-func`) that allows the compiler to insert inaccurate approximations for the functions found in Chapter 7. These approximations are often rough versions that are off by several ULPs compared to the implementations in the standard math library. In both cases, the compilers will also not be shy about fusing an addition and a multiplication into an FMA operation, but this is the default compiler behavior unless it is explicitly disabled via the `-ffp-contract` flag.

If you do plan to use these flags, it is preferable to also use a flag to set the target machine architecture, because the performance of some of these optimizations is worse on more recent architectures than the unoptimized code. For example, the performance of division in floating point units has been improving rapidly, so there are many places in which a CPU from 2024 or later would prefer to do a floating-point division while a CPU from 2014 would benefit from math on a reciprocal. Compilers will also sometimes not compute this reciprocal to full machine precision, leaving you with an inaccurate result that is also slower than the "unoptimized" code. Fast math compiler options compromise correctness, so make sure that tradeoff is worth it!

Additionally, there are a few processor settings that can be enabled to balance performance and accuracy at the hardware level. These flags mostly control behavior around subnormal numbers. Since older hardware will call microcode to deal with subnormal numbers, and this behavior is slow, many processors have a setting that allows you to have the processor treat subnormal numbers as zero, and to flush any subnormal outputs of functions to zero. These settings are often called **denorms are zero** for instructions to treat subnormal inputs as zero and **flush to zero** to flush subnormal outputs to zero. These options help to prevent non-determinism in machine instructions and keep your code away from slow fallback paths in software. However, hardware has gotten a lot better at handling subnormal numbers.

If you have the time to work out your code's performance, it is often better to avoid instability by just doing the relevant optimizations for yourself and avoiding the compiler's meddling. Options that allow aggressive optimization will compromise everything they touch from a numerical perspective, when you would likely prefer to be a bit more surgical. In calculations for which you would turn on the fast math optimizations, you can often accept *far more* error than fast math will give you, and if you can use the techniques discussed in later chapters to make the speed-error tradeoff for yourself, you will do significantly better than the compiler while being explicit about how much error you decide to accept. Of course, this comes at the cost of mental effort, which also must be balanced in the equation.

Compiler Settings for Fast, Safe Math

Most of the code you write in a simulator or game will be executing a lot, so it helps to allow your compiler *some* flexibility to optimize your mathematical code in your production builds without going quite as far as the full fast math setting. The settings listed below are recommended to give your compiler maximum flexibility without compromising your math too significantly.

- `-fno-signaling-nans` and avoid using signaling NaNs.

- `-fno-trapping-math` and `-fno-math-errno` to disable anything but explicit error and exception handling in your optimized builds.

- `-fno-rounding-math` in the 99% of cases that do not change rounding modes (this is actually a compiler default).

Options like the components of `-funsafe-math-optimizations` and `-fexcess-precision=fast` can have significant numerical effects. The associative math flag in particular is known to be a numerical bugbear that changes your math significantly. On the other side, `-fno-signed-zeros` is a candidate for this list if you aren't going to be relying much on the behavior of $\frac{1}{0}$ or similar operations that need the information given by a signed zero.

Check Your Understanding

Problem 9.1. Find the most accurate form of the quadratic formula when you assume a, b, and c are all positive and of similar magnitude $\leq 2^{50}$.

Problem 9.2. Simplify Algorithm 9.4 to take the product of a double-double number and a double-precision number.

Problem 9.3. Heron's rule finds the area of a triangle from its side lengths. Using a triangle of side lengths a, b, and c, Heron's rule tells us that the area of the triangle is:

$$A = \sqrt{s(s-a)(s-b)(s-c)}$$

where s is half the perimeter:

$$s = \frac{a + b + c}{2}$$

Perform a backward error analysis of Heron's rule in a (b and c are symmetric). What is the backwards error with a relatively large (up to $a \approx b + c$)? What is the backwards error with a relatively small ($b \approx c$)?

Problem 9.4. Return to problem 4.3 with Algorithm 9.6. How much more accurate is the result?

Problem 9.5. Devise a floating-point calculation that varies by more than 1% between a program compiled without fast math and a program compiled with fast math turned on.

Problem 9.6. Come up with an algorithm for accurately computing the volume-weighted average price of a stock over a month. The volume-weighted average price is given by:

$$\text{VWAP} = \frac{\sum s_i p_i}{\sum s_i}$$

where s_i is the number of shares traded on each trade, and p_i is the price of that trade. Over a month, an expected number of 100 million shares trade hands in popular stock symbols on the US stock exchanges, and each trade averages about 100 shares at a price between \$5.00 and \$500.00.

Bonus: Redesign your calculation for any popular cryptocurrency and check the accuracy of an online source.

Problem 9.7. Extend the compensated summation algorithm (Algorithm 9.6) to use a quad-double sum instead of a double-double sum. Find a list of numbers that causes Algorithm 9.6 to have significant error while your extended-precision version is correct. Find a list of numbers on which your extended-precision version still fails.

10

Function Approximation

Function approximation, for many users of math libraries, is like assembly language. Understanding how it works helps you make intelligent decisions at a higher level, although you will rarely write your own math libraries. As discussed before, many platforms use their own math libraries, and this can introduce subtle errors between platforms. The available math libraries are often engineered to balance speed and precision for a given platform, language design philosophy, and application space.

On one end of the spectrum, it is possible to construct very precise approximations that are within half an ULP of the real-valued version of a function even when using tens of floating point operations to do it. On the other side, it is also possible to produce very fast approximations with relatively few bits of accuracy. The former is usually left to math libraries for common functions, but if you leave this up to dynamic linking, you may be faced with libraries of varying quality. That can mean that calculations are not repeatable between machines, and some operating systems and platforms may deliver undesirable results. The latter is generally the domain of custom handwritten math libraries.

For this reason, it is useful to understand and appreciate function approximation and the tradeoffs that you can make when selecting math libraries, and if you find yourself spending a lot of computer time on one single function, there is usually a nice spot on the speed–accuracy tradeoff curve that you can find with a bit of human time.

10.1 Accuracy of Approximations

We can look at accuracy of approximations using a number of different metrics. Generally, functions are approximated over a range of values of interest. Over that range, the most common measure of accuracy to optimize is the maximum error of the function over that range. Expressed mathematically, the error of an approximation $\hat{f}(x)$ against $f(x)$ is computed as:

$$E_{\text{minimax}} = \max \left| f(x) - \hat{f}(x) \right|$$

DOI: 10.1201/9781003565543-10

By minimizing this measure of error, we are minimizing the maximum error, giving this form of problem its name: A "minimax" problem. Our goal is to minimize the maximum error. This metric does not care about average error, only the error at the specific point(s) where it reaches its worst value. This is also referred to as the $\mathcal{L}^\infty$ norm in mathematical contexts. Minimax error is often the default objective function when you intend to produce an approximation that is very close to your target function. For this reason, several methods require the use of minimax error [24].

It is also possible to consider the root-mean-squared error (RMS error, also $\mathcal{L}^2$ norm), although this is not common for function approximations:

$$E_{\text{RMS}} = \sqrt{\int_\alpha^\beta \left[f(x) - \hat{f}(x) \right]^2 dx}$$

where α and β are the bounds of the range of interest. Root-mean-squared error penalizes your function for being slightly wrong at certain points, with increasing penalty for large errors. Unlike minimax error, the RMS error metric does pay attention to every point on the function, but it is often harder to work with when doing function approximations, since you have to deal with the integral in the square root.

For approximations that cover a widely varying range, it can be more helpful to consider the accuracy of the approximation relative to the value of the function. An absolute error of 0.1 has a very different meaning when a function's value is 0.0001 than it does when a function's value is 100. Relative accuracy is often measured in bits:

$$A_b = -\max \left(\log_2 \left| \frac{f(x) - \hat{f}(x)}{f(x)} \right| \right)$$

Applying this formula, a function whose value is 100 with absolute error of 0.1 is accurate to 9.96 bits. In many cases, this metric of accuracy lines up better with the idea of floating point than metrics of absolute error, although approximations that operate only on the mantissa bits of a number can generalize well to a wide range when constructed using minimax error. However, this metric of accuracy can give useless numbers when $f(x)$ is near zero if you really care about absolute error, as even a small absolute error in that range results in a large relative error. It is also possible to remove the log from the accuracy format and measure accuracy as a fraction or percentage of error.

An additional wrinkle to deal with in floating point is that your number space is not continuous. Some baseline error that comes from using a floating point number system, related to the quantization of the number space. This quantization can also cause some additional approximation error, since constants and coefficients also have to be rounded to fit into floating-point numbers, and each internal partial result is also rounded. For this reason, the optimal coefficients for a floating-point approximation are often subtly different than the optimal coefficients for an approximation using the real numbers.

A function that is accurate to the full width of a floating-point format and rounded to the right value for a given rounding mode is called a "correctly rounded" function. Correctly rounded functions return the same result as if you had computed the function at infinite precision and then converted to floating point at the end, and have an error of half a ULP using default rounding. A weaker term, "faithful rounding" refers to functions that produce a correct result within one ULP. Many math libraries stop at faithful rounding, but recent advances are making correctly rounded math libraries possible.

10.2   Polynomial Approximations

We have a few basic arithmetic operations that are capable of computing polynomials, but cannot compute transcendental functions like $\exp(x)$ or $\sin(x)$ directly. A logical solution to this problem is to construct a polynomial that approximates a target transcendental function.

Polynomial approximations of transcendental functions have been studied since the Taylor series. A Taylor series is often a poor approximation, because it is defined by the behavior of a function only at a specific point ($x = 0$ in this example):

$$f(x) = f(0) + f'(0)x + \frac{f''(0)x^2}{2} + \frac{f'''(0)x^3}{6} + \cdots = \sum_{i=0}^{\infty} \frac{f^{(n)}(0)x^n}{n!}$$

Going to order infinity, the Taylor series of f perfectly matches f, but essentially categorizes its behavior based on its curvature around one point.

If you cut off the Taylor series at the nth coefficient, you get a polynomial that approximates a function f with error of order n. However, the order of the error says nothing about the constant on that error, and Taylor series are prone to diverge for functions that are infinitely differentiable. Notably, most functions of interest to approximate are infinitely differentiable. At the exact chosen point, the Taylor series is a phenomenal approximation, but as we move away from it, the approximation gets worse. Ideally, we would like our approximation to have roughly equal error in a region of interest rather than error that grows as we move away from a point. As an example, the Taylor series of $\frac{1}{x}$ about $x = 1$ is shown in Figure 10.1.

The Taylor series is an alternative way to express a function that gives you a polynomial representation. In mathematical terms, the Taylor series expansion is actually a change of basis for a function (a basis is a representation space, and a change of basis is akin to a generalization of the idea of a change of coordinates). Another change of basis for a function that you might be

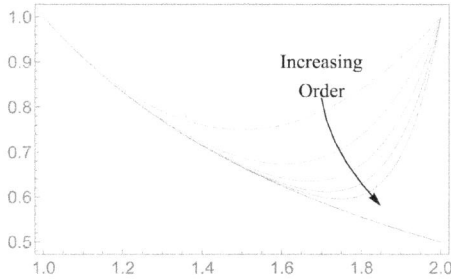

FIGURE 10.1
Even-order Taylor series from order 2 to order 10 of $\frac{1}{x}$ about $x = 1$, plotted in the region between 1 and 2 against the actual function (in black). Although higher-order approximations stay near $\frac{1}{x}$ for longer, they all blow up as $x \to 2$, with higher-order approximations blowing up faster.

familiar with is the Fourier series:

$$f(x) = a_0 + \sum_{n=1}^{\infty} a_1 \cos(nx) + \sum_{n=1}^{\infty} b_1 \sin(nx)$$

For a Fourier series, we don't think of the function in terms of powers of x any more, but in terms of waves of $\cos(nx)$ and $\sin(nx)$. Like a Taylor series, a Fourier series can break down a function into different constituent parts. A Taylor series breaks down a function by its curvature, and a Fourier series breaks down a function based on its oscillatory components. The Fourier series is also relatively useless for approximation, both because it needs a high order to closely approximate a function, and also because it requires a trigonometric function to compute.

However, a less-known relative of the Fourier series is very useful for numerical approximations, namely the **Chebyshev series**, which is built on windowed polynomials first described by Russian mathematician Pafnuty Chebyshev. Before going on to describe them mathematically, a plot of the first 6 Chebyshev polynomials is shown in Figure 10.2. In the region $[-1, 1]$, the Chebyshev polynomial of order n crosses zero n times, and touches either $y = 1$ or $y = -1$ a total of $(n + 1)$ times. In other words, the Chebyshev polynomials are a generic type of minimax polynomial in this region. They are also the polynomials with the highest leading coefficient that stay in this region, and they have a number of other very neat mathematical properties. All of this leads to the fact that the Chebyshev polynomials in the region $[-1, 1]$ are another way to decompose a function in that region [33].

Mathematically, the Chebyshev polynomials are solutions to the equation:

$$T_n(\cos(\theta)) = \cos(n\theta)$$

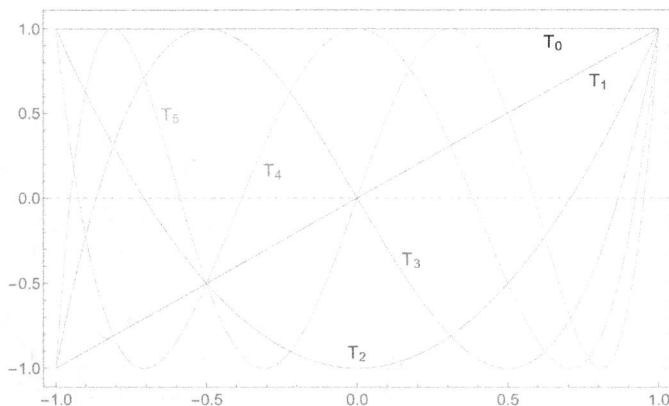

FIGURE 10.2
Plot of Chebyshev polynomials of order 0–5. Each polynomial crosses $y = 0$ a number of times equal to its order, and touches either $y = 1$ or $y = -1$ a number of times equal to its order plus one.

Or they can be expressed by the recursion:

$$T_0(x) = 1$$
$$T_1(x) = x$$
$$T_{n+1}(x) = 2xT_n(x) - T_{n-1}(x)$$

In turn, an function in the region $[-1, 1]$ can be expressed as a sum of Chebyshev polynomials, with coefficients computed similarly to Fourier series coefficients:

$$f(x) = \tfrac{1}{2}c_0T_0(x) + \sum_{n=0}^{\infty} c_nT_n(x)$$

For completeness, it is possible to calculate the c_n terms by doing a Chebyshev transform directly using the following integral:

$$c_n = \frac{2}{\pi} \int_{-1}^{1} \frac{f(x)T_n(x)}{\sqrt{1 - x^2}} dx$$

Chebyshev approximations of functions are rarely computed this way, however, as this is usually harder than other methods. Still, this formula can be useful for high-order approximations, where other approaches become more difficult.

The method of Chebyshev approximation expands to arbitrary regions by mapping the target region into the region $[-1, 1]$. When the Chebyshev series of f is cut off, you get a polynomial approximation of f that is exactly equal to f at the zeros of that Chebyshev polynomial. For this reason, the zeros of the Chebyshev polynomial when mapped into a region are often called "Chebyshev points" or "Chebyshev nodes", and we actually calculate Chebyshev approximations by solving an equation.

In the region $[-1, 1]$, the zeros of the Chebyshev polynomial (**Chebyshev zeros**) of order n can be found at:

$$x_k = \cos\left(\frac{k - \frac{1}{2}}{n}\pi\right) \qquad (k = 1, 2, \ldots, n)$$

The extrema of the Chebyshev polynomials are also called "Chebyshev nodes" sometimes, but we will refer to them as **Chebyshev extrema**. They are at the x values:

$$x_k = \cos\left(\frac{k}{n}\pi\right) \qquad (k = 0, 1, 2, \ldots, n) \qquad (10.1)$$

We can get a good order-N approximation of a function by solving for a polynomial where f equals that polynomial at the Chebyshev zeros. In equation form, we can create a polynomial approximation of f by solving for the a's in the system of equations given by:

$$f(x_k) = \sum_{n=0}^{N} a_n x_k^n \quad \textbf{for} \quad x_k = \cos\left(\frac{k - \frac{1}{2}}{n}\pi\right) \quad \textbf{when} \quad (k = 1, 2, \ldots, N+1)$$

This polynomial will be equal to the polynomial gathered by computing a Chebyshev series and truncating due to another fascinating property of the Chebyshev polynomials: The sum of all Chebyshev polynomials of degree greater than n is 0 at the Chebyshev zeros of order n, so a Chebyshev approximation to order $(n-1)$ is exact at those values of x.

As an example of this process, we will construct a third-order approximation for $\sin(x)$. We will begin by constraining the window. As we can see from Figure 10.3, we only need to approximate the region $\left[0, \frac{\pi}{2}\right]$, since sine is periodic, and that period is constructed out of four windows that are symmetric to the behavior in this region, so we can extend our approximation to work for any input with only a quarter of a period. We first map the Chebyshev zeros from the region $[-1, 1]$ to the region $\left[0, \frac{\pi}{2}\right]$:

$$x_k = \frac{\pi}{4}\left[\cos\left(\frac{k - \frac{1}{2}}{n}\pi\right) + 1\right] = \{1.51101, 1.08596, 0.484839, 0.0597849\}$$

Then we solve:

$$\sin(1.51101) = a_0 + a_1(1.51101) + a_2(1.51101)^2 + a_3(1.51101)^3$$
$$\sin(1.08596) = a_0 + a_1(1.08596) + a_2(1.08596)^2 + a_3(1.08596)^3$$
$$\sin(0.484839) = a_0 + a_1(0.484839) + a_2(0.484839)^2 + a_3(0.484839)^3$$
$$\sin(0.0597849) = a_0 + a_1(0.0597849) + a_2(0.0597849)^2 + a_3(0.0597849)^3$$

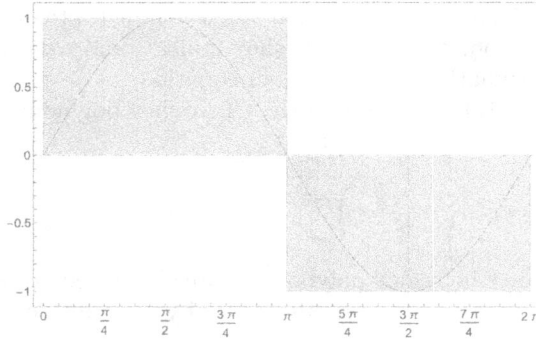

FIGURE 10.3
Plot of one full period of sin(x) showing the windows in which the function is symmetric. Every gray box has the same curvature, but is flipped in the x and y directions.

The resulting equation is:

$$\sin(x) \approx -0.00113088 + 1.0227x - 0.0664781x^2 - 0.114263x^3$$

Graphically, this generates an approximation that is shown in Figure 10.4, which is a lot closer than the Taylor series. Unlike the Taylor series, Chebyshev approximations do not have divergence problems.

However, while a Chebyshev approximation will be close to minimax error, the error is somewhat imbalanced throughout the window. Looking at Figure 10.4 again, we can see that the error is larger when $f(x)$ is larger. This is a non-optimality of Chebyshev approximations, but if we want to balance the error to truly minimize the minimax error, we need another trick.

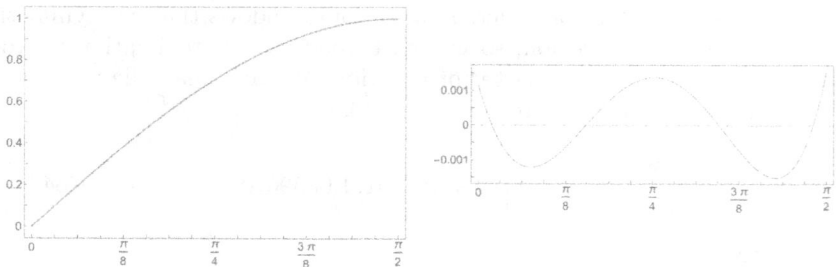

FIGURE 10.4
Left: Third order Chebyshev approximation of sin(x) (black dashed line) compared against the function (gray). The approximation lines up nearly perfectly. Right: Error of the approximation.

That trick is the **Remez exchange** algorithm. The first step of the Remez algorithm is similar to the construction of a Chebyshev approximation, but adds an extra variable to the system of equations to solve representing an alternating error term, E, and solves that system at the Chebyshev extrema rather than at the Chebyshev zeros. In other words, instead of solving for the places where the approximation is exactly equal to the function, we try to solve for the locations of maximum error. This makes intuitive sense given that we care more about maximum error than exactness [34].

Starting with the set of x_k given by Equation 10.1, we solve the system:

$$f(x_k) = \sum_{n=0}^{N} a_n x_k^n + (-1)^k E \qquad (k = 0, 1, \ldots, N+1) \qquad (10.2)$$

Solving at the extrema means that we use $N + 2$ equations to solve for an approximation of order N, but we also simultaneously solve for a minimax error at each Chebyshev extremum. The effect of this is that we balance the error at the Chebyshev extrema, but in exchange for that guarantee, this initial approximation has no guarantee that the Chebyshev extrema will be the local maxima of the error. Remez's insight is that we can refine our approximation by doing the following:

1. Find the actual local maxima of error in our approximation.

2. Replace the set of x_k with the x values of the local maxima.

3. Solve the system of equations given in Equation 10.2 again to get new a_n and E.

4. Repeat from (1) until convergence.

For $\sin(x)$ at third order, the resulting equation (converging in double precision after only one round) is:

$$\sin(x) \approx -0.00135867 + 1.02523x - 0.0706971x^2 - 0.112491x^3$$

A comparison of error for our Remez approximation against the Chebyshev approximation from before can be found in Figure 10.5. The error from the Remez approximation is larger early in the range, but is smaller later. By balancing error, the maximum error is reduced compared to the unbalanced Chebyshev approximation. However, if we are interested in a different metric of accuracy than minimax error, a Remez approximation may be suboptimal.

The Chebyshev and Remez approximations are a starting point for many polynomial approximations used today. Like we saw with reciprocal, sine, and square root, most functions have a small window in which we need to actually approximate, due to periodicity and exponent-related algebra. However, translating from infinite-precision coefficients to coefficients that work in floating point is a separate undertaking. All of the coefficients from Chebyshev and Remez approximations are heavily interrelated, and so an accurate quantization cannot consider them separately. A proper quantization needs to consider

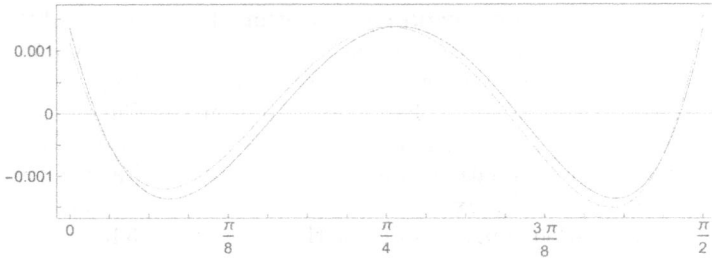

FIGURE 10.5
Error of Remez (black) and Chebyshev (gray) approximations of $\sin(x)$. The Remez error is more balanced across the domain.

that when one coefficient is rounded up, another coefficient will likely need to be rounded down to compensate.

Floating-Point Aware Polynomial Approximation

A simplistic approach to quantizing the coefficients of a polynomial approximation is to look at the space of all floating point numbers near each coefficient and pick the combination that is the most accurate. This can mean considering a few hundred or a few thousand possibilities.

More recently, numerical analysts have started to work directly with the bits of floating-point numbers instead of producing floating point approximations by computing in infinite precision and then quantizing. The naive approach of quantizing coefficients separately can end up rounding several coefficients in the same direction, which introduces error into an approximation. Similarly, the optimal approximation when coefficients are quantized and rounding occurs between operations can be far enough from the Remez coefficients that you will not be able to find them by simply rounding each Remez coefficient in both directions and seeing which ones are the best [35].

Two types of solvers are used for these approximations, either linear programming (LP) solvers, which can solve generic linear optimization problems in a quantized number space (with great effort), or solvers based on integer relation algorithms. Integer relation algorithms, such as the Lenstra–Lenstra–Lovász (LLL) lattice basis reduction algorithm, find "nice" quantizations for sets of real numbers while maintaining a relation between them. This class of algorithms is not NP-hard. In the linear programming approach, the solver is given a bit more freedom at a cost of compute and algorithm complexity, while the LLL approach quantizes something that has been pre-made.

Tools like `sollya` allow you to use these algorithms without having to program them. Sollya's `fpminimax` computes polynomial approximations by starting with Remez, and then quantizing coefficients with LLL. This method allows the creation of coefficient sets that give precise results without doing too many extended-precision operations that cost efficiency [36]. This takes the guesswork out of quantization.

There are several LP solvers that can be programmed to work on floating point numbers with great difficulty. As a practical matter, some solvers are also better at certain problems than other solvers, and this means that it is tougher to take a generic approach to solver-based quantizations.

Given the problem of approximating a function today, the following approach is suggested:

1. **Range Restriction:** Look for periodic behavior or fractal behavior in a function or in parts of a function to find the smallest possible range under which function evaluations can be restricted.

2. **Approximation:** Use a tool like `sollya` to construct a minimax approximation of the restricted function on that range.

3. **Function Wrapping:** Handle special cases, including invalid cases and infinities, as needed, using separate logic in a wrapper function around the range-restricted approximant.

10.3 Newton-Raphson and Iterative Methods

Functional iteration methods are a complement to polynomial approximations. In general, a functional iteration method takes a "guess" x_0 that is close to some target, and refines it into a better guess using a crafted iteration function. Successive uses of an iterative method will converge toward a correct result, often quickly. As we saw in Section 4.2, some functions that are hard to evaluate can be reformulated as refinement iterations that are easy to compute.

Newton-Raphson iteration (also called Newton's method) is the basic form of this. Theoretically, Newton-Raphson finds zeros of a function f based on an initial guess, x_0. We then refine that guess by walking down the slope of f:

$$x_{i+1} = x_i - \frac{f(x_i)}{f'(x_i)}$$

When we reach a value of x for which $f(x) = 0$, we stop there automatically, but until we get there, we step toward the nearest zero of f. In itself, this is not necessarily useful unless we can craft f so that its zeros are meaningful, and start with x_0 close enough to that zero that we have a chance to find it. For example, to compute $\frac{1}{D}$, we can choose:

$$f(x) = \frac{1}{x} - D$$

This function only reaches zero at $x = \frac{1}{D}$. The Newton-Raphson iteration is then:

$$x_{i+1} = x_i - \frac{\frac{1}{x_i} - D}{\frac{-1}{x_i^2}} = x_i + x_i - Dx_i^2 = x_i(2 - Dx_i)$$

This form of iteration allows us to turn a reciprocal calculation into a sequence of steps where each step needs a multiplication and an FMA.

A famous example of Newton's method is the Babylonian square root, which turns a square root into an iterative algorithm of the four basic arithmetic operations, using $f(x) = x^2 - a$:

$$x_{i+1} = x_i - \frac{x_i^2 - A}{2x_i} = x_i - \left(\frac{x_i}{2} - \frac{A}{2x_i}\right) = \frac{1}{2}\left(x_i + \frac{A}{x_i}\right)$$

However, Newton's method is most often useful for reciprocals and reciprocal roots, and the function $f(x)$ must be carefully chosen to avoid divergence and oscillation. Calculation of square root by Newton-Raphson usually involves calculating inverse square root and then multiplying it by the original number.

A typical example of divergence of Newton's method, although an impractical one, is $f(x) = \sqrt[3]{x}$. This function obviously has a zero only at $x = 0$, but the Newton-Raphson iteration looks like:

$$x_{i+1} = x_i - \frac{x_i^{1/3}}{\frac{1}{3}x_i^{-2/3}} = x_i - 3x_i = -2x_i$$

If you start iterating, even arbitrarily close to $x = 0$, this iteration will diverge. This is a pathological example, but it illustrates that the choice of f must be deliberate to get good results. A final issue that limits the effectiveness of this method is that we often need to compute transcendental functions to use it. For example, we can attempt to apply Newton's method to compute natural logarithms by calculating with $f(x) = \exp(x) - A$:

$$x_{i+1} = x_i - \frac{\exp(x_i) - A}{\exp(x_i)} = x_i - 1 + \frac{A}{\exp(x_i)}$$

This function should converge to $x = \ln(a)$, but we need to compute $\exp(x_i)$ to get there, and that takes as much computing power as computing the log directly *for each iteration*. We will only be able to compute $\ln(x)$ this way if we are willing to wait, and an approximation of $\exp(x_i)$ will not work. You may have noticed that we have chosen f so that there is always only one zero of the function and that the arithmetic cancels nicely. That is not an accident—that is by design—and to use Newton-Raphson, you have to design things similarly.

When Newton-Raphson does work, however, it works fast. When it converges, Newton-Raphson has *quadratic convergence*, meaning that it squares the error at each step. This practically means converging at an exponential rate, doubling the number of correct bits in your result at each iteration. To show the convergence speed, we can consider the error term for step i for some $f(x)$ that has a zero at $x = \xi$:

$$\epsilon_{i+1} = x_{i+1} - \xi$$

We can abbreviate the Newton-Raphson iteration to g:

$$x_{i+1} = g(x_i) = x_i - \frac{f(x_i)}{f'(x_i)}$$

Since we know that Newton-Raphson is stationary at $x = \xi$, we can say that:

$$x_{i+1} - \xi = g(x_i) - g(\xi)$$

We can take a Taylor series expansion of g about ξ, stopping with a second-order remainder (R_2):

$$x_{i+1} - \xi = g(\xi) + g'(\xi)(x_i - \xi) + R_2 - g(\xi)$$

Taylor's theorem indicates that there is a constant c for which $R_2 = c(x_i - \xi)^2$:

$$x_{i+1} - \xi = g(\xi) + g'(\xi)(x_i - \xi) + c(x_i - \xi)^2 - g(\xi)$$

We can also expand the derivative of g, which cancels nicely:

$$g'(x) = 1 - \frac{f'(x)^2 - f(x)f''(x)}{f'(x)^2} = \frac{f(x)f''(x)}{f'(x)^2}$$

That means that since $f(\xi) = 0$, $g'(\xi) = 0$. This should not be surprising given that the iteration is stationary at $x = \xi$. So with the $g(\xi)$ terms canceling, and the $g'(\xi)$ term going to zero, we are left with:

$$x_{i+1} - \xi = c(x_i - \xi)^2$$
$$\epsilon_{i+1} = c\epsilon_i^2$$

Within a constant factor, we square our error at each iteration. That constant is usually 1 for the useful Newton-Raphson methods.

However, Newton-Raphson iterations can still be affected by rounding and quantization error, which can cause iterative approximations to reach an accuracy floor before getting to the level of being correctly rounded [37]. Consider the division iteration:

$$x_{i+1} = x_i(2 - Dx_i)$$

Generally, when we have a good guess at $x_i = \frac{1}{D}$, Dx_i will be very close to 1, meaning that $2 - Dx_i$ will also be very close to 1. That means an effective loss of precision for this part of the calculation since the implied bit of the significand is far away from the bits where the action is, leading to a number of leading zeros in the mantissa. A much more precise number is the error of Dx_i, which is very close to 0:

$$E_i = 1 - Dx_i$$

That means that we can likely get a better approximation by first computing the error, and then using an FMA operation to update x_i:

$$x_{i+1} = x_i + E_i x_i$$

Instead of using an FMA to get an imprecise $2 - Dx_i$ and multiplying it by x_i (where both operations are nearly guaranteed to be inexact), we can use a pair of FMAs to keep more bits of precision through the calculation. This is another example of doing algebra to keep calculations precise.

There are higher-order methods with faster convergence than Newton-Raphson, but they are rarely used in practice. **Halley's method** (the same Halley as the comet) is the second-order version of Newton-Raphson, which has cubic convergence, tripling the number of correct bits at each iteration:

$$x_{i+1} = x_i - \frac{2f(x_i)f'(x_i)}{2f'(x_i)^2 - f(x_i)f''(x_i)}$$

This iteration is much more complicated than the Newton-Raphson iteration step and will usually take more operations unless f is very carefully chosen, but if you can find a way to make the hard parts cancel out, it will converge quickly.

Newton-Raphson and its higher-order variants have one final advantage in that they self-correct for quantization error. Quantization can result in a Newton-Raphson iteration being slightly larger or slightly smaller than expected, but when that occurs, the next iteration takes the actual value computed at the previous round into account. In that sense, there is no translation problem for Newton-Raphson methods when moving from a method made in the land of the real numbers to a method made in floating point as long as any constants in the iteration can be represented exactly.

Newton-Raphson takes some care to deploy, and in the context of function approximation, that usually means crafting a very simple function that nicely cancels out transcendental parts of function calculations or other "hard" parts. However, when it is able to be deployed, it can converge to very accurate results quickly, with the only limits on its accuracy coming from the precision of the arithmetic used to compute the iteration. Polynomial approximations can take a long time to compute to reach a correctly rounded result, while a combination of a Remez approximation for an initial guess with a well-designed Newton-Raphson iteration can get there more efficiently. The hardest part is finding the iteration algorithm.

10.4   Lookup Tables and Interpolation

A tempting alternative to function approximation is to use lookup tables (LUTs). However, it is usually not the fastest solution for a given accuracy level unless you have hardware that is built to support your lookup table approach (e.g., on a GPU) or construct your lookup table with some intelligence. Some situations exist where you have a function with a limited input space, extreme nonlinearities, and a difficult computation path. In these circumstances, the use of lookup tables can be the fastest way to build an algorithm.

The core idea of using a lookup table to compute a function is that you are turning an arithmetic problem into a memory lookup. This means that the exact computing platform being used can affect whether lookup tables are a good idea. In general, GPUs are built to support lookup tables and CPUs are terrible at computing them. With floating-point inputs, any lookup table has to cover the entire dynamic range of input while maintaining high precision— even restricting the domain of the table to $[0, 1]$ means accepting a quarter of all possible floating point numbers. Practically, this means using some form of interpolation.

On CPUs, function approximations based on lookup tables are usually undesirable, since the CPU cache hierarchy can cause table lookups to have performance that is highly variable. Additionally, the use of lookup tables causes you to consume a limited global resource, memory bandwidth, to perform a computation that could otherwise be done on a core. It also adds variance to the performance of a calculation, which is often undesirable in itself. As an example, cache and memory access latencies for several CPUs are shown in Table 10.1. As a result, lookup tables on CPUs can be a dubious idea, especially when you don't control the other programs running on the machine or its design.

Thanks to the cache hierarchy, the performance of lookup tables varies greatly depending on the frequency at which the function is called. An infrequently called function or a large lookup table will very often fetch from main memory or from far down in the cache hierarchy, causing both a local slowdown while the memory fetch occurs and a global slowdown since you are hogging shared resources. Conversely, small lookup tables that are used frequently can often reside in core-local caches with plenty of available bandwidth and fast access times. Lookup table size has a pseudo-quadratic effect on speed. When your lookup table is larger, it occupies more room in a cache and each cache line of the lookup table is touched less frequently since you will only look up one entry per use of the table. Finally, the cache hierarchy makes lookup tables look good in benchmarks. A microbenchmark does nothing but call a single function, which leads to your lookup table staying as close to the CPU as possible, while in a real program, it will often be further down the cache hierarchy than you want.

TABLE 10.1

Cache and memory sizes and access latency on the AMD Zen 4 client architecture and the P-Core on Intel's Meteor Lake client platform. The Zen 4 device was tested with DDR5 memory, while the Meteor Lake device had LPDDR5x memory, another source of variance.

Memory Layer	Zen 4 [38]		Meteor Lake [39]	
	Size	Latency	Size	Latency
Caches				
L1 Data Cache (local)	32 kB	0.7 ns	48 kB	1 ns
L2 Cache (local)	1 MB	2.5 ns	2 MB	3.3 ns
L3 Cache (global)	4 MB/core	8–9 ns	3 MB/core	15 ns
Main Memory				
DDR5	Large	75 ns		
LPDDR5x			Large	170 ns

In the best case, a small lookup table that is accessed frequently can be accessed in 4 CPU clock cycles, during which time a modern CPU can perform 16 or more vector arithmetic instructions. In the worst case, larger lookup tables can occupy the compute time of more than 5 000 individual FMAs. This leads to a few guidelines as to when CPU lookup tables are useful as approximators [40]:

1. When a lookup table is one cache line, it is equivalent to accessing a variable or coefficient.

2. When a *full-system* benchmark indicates that the lookup table is faster than the calculation you would otherwise do, use it.

3. When you cannot come up with any other way to evaluate the function, use a lookup table.

On GPUs, there is usually a section of fast memory reserved for constants and lookup tables. These caches are also often used for textures, and can be quite large. They also are used for an important part of a rendering pipeline: Color grading. In rendering and video editing, color grading can involve very difficult nonlinear functions that have to be evaluated for every pixel and are often driven by artistic taste, so it helps to use a relatively small lookup table that is precomputed once, mapping from an input color to an output color. Using a lookup table mapping also allows artists to "draw" this mapping function instead of trying to construct something numerically.

Finally, if you happen to be building hardware or working with other exotic computing devices, lookup tables are often the best way to evaluate functions. For hardware, ROM is relatively cheap, and there are many ways that you can compress a lookup table to make it fit in a smaller memory

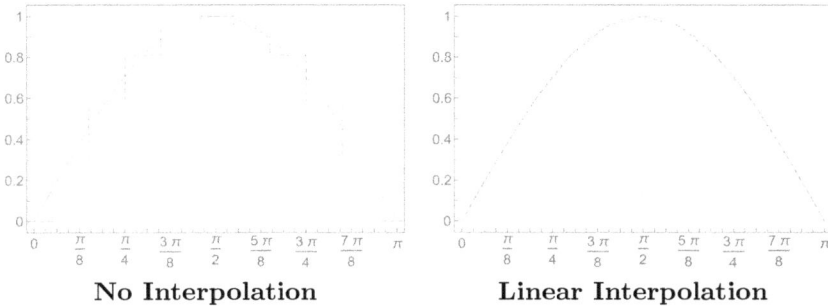

No Interpolation   Linear Interpolation

FIGURE 10.6
Comparison of a lookup table with no interpolation, taking the value of the nearest lookup table point, against linear interpolation of a lookup table for $f(x) = \sin(x)$, where both plots show the approximation (dashed black line) plotted against the function (gray line).

or circuit area, usually using arithmetic operations to combine several small lookup tables [41].

Using lookup tables with floating point values goes hand-in-hand with interpolation. Floating point values will often lie between the cells of a lookup table, so interpolation allows you to construct a synthetic value from the neighboring points on the lookup table. A visual example of linear interpolation is shown in Figure 10.6.

The most common form of interpolation, linear interpolation (`lerp`), is done by taking a weighted average of the two nearest lookup table outputs, weighted by the distance to each one. Taking that distance as a parameter (t) where t is normalized so that $0 \le t \le 1$, linear interpolation computes:

$$\text{lerp}(a, b, t) = a(1 - t) + bt$$

For lookup table entries $(x_0, f(x_0))$ and $(x_1, f(x_1))$, given an input x, the formula to linearly interpolate $f(x)$ incorporates that normalization:

$$f(x) \approx \frac{f(x_0)(x - x_0) + f(x_1)(x_1 - x)}{x_1 - x_0} = f(x_0) + \frac{f(x_1) - f(x_0)}{x_1 - x_0}(x - x_0)$$

This allows you to smear your way between two y values smoothly, but does not account for any curvature in the function you may be approximating. Linear interpolation also has many uses outside of lookup table approximations, since it is a general way to smooth out the transition between two points.

For smooth functions, the error of linear interpolations is bounded by the granularity of the lookup table and the amount of curvature of the function (i.e., the function's second derivative). As expected, linear interpolation is a first-order approximator, and has second-order error. Our error is:

$$E(x) = f(x) - f(x_0) - \frac{f(x_1) - f(x_0)}{x_1 - x_0}(x - x_0)$$

There is some point $x = \xi$ between x_0 and x_1 at which we reach the worst error in the window. This point can be a local minimum for E or a local maximum depending on whether we overestimate or underestimate by linearly interpolating. If we take a Taylor series expansion of E around ξ with second-order error, we get:

$$E(x) = E(\xi) + E'(\xi)(x - \xi) + c(x - \xi)^2$$

Since $E(\xi)$ is the point of the worst error, $E'(\xi) = 0$:

$$E(x) = E(\xi) + c(x - \xi)^2$$

We know the value of $E(x)$ at a few values of x. Specifically, we know that $E(x_0) = 0$. Plugging x_0 into this formula, we get:

$$E(x_0) = 0 = E(\xi) + c(x_0 - \xi)^2$$

So our maximum error is proportional to $(x_0 - \xi)^2$:

$$E(\xi) = -c(x_0 - \xi)^2$$

Thus, we have second-order error when we use linear interpolation as a method of approximation.

When you have functions that are either concave or convex—where the curvature of the function only goes one way—the error of a lookup table constructed out of points $y = f(x)$ will be unbalanced. For a convex function, a lookup table with linear interpolation will always overestimate, while for a concave function, a lookup table will always underestimate. it is possible to have better accuracy by constructing the table in an error-aware way. Instead of sampling $f(x)$ exactly for each lookup table point, it is possible to balance error on these functions by moving the corresponding lookup table points up or down. With this correction, lookup table error can be halved.

It is also useful to use lookup tables with higher-order approximations in each table row, and lookup tables can be used for domain separation of inputs. That way, lookup tables can be used in combination with minimax approximations, like Remez approximations, to get the best of both worlds. This can also be done with higher-order interpolation algorithms such as spline interpolation. A low-order polynomial approximation in each cell of a small lookup table can produce a good compromise of performance and accuracy.

It is possible to extend linear interpolation into more dimensions by inter-polating in one dimension at a time, interpolating between the results at each step. For example, if you have a two-dimensional input, we now take the near-est four table entries and interpolate first in the x dimension to get two extreme values of $f(x, y_0)$ and $f(x, y_1)$, then we interpolate between those values. This is called **bilinear interpolation**, and a graphical representation of what is happening is shown in Figure 10.7. Mathematically, with lookup table entries

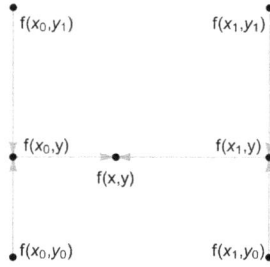

FIGURE 10.7
Graphical view of bilinear interpolation. We start with four sample points from our lookup table, then linearly interpolate in one dimension to get two, then linearly interpolate the final point.

$(x_0, y_0, f(x_0, y_0))$, $(x_1, y_0, f(x_1, y_0))$, $(x_0, y_1, f(x_0, y_1))$, and $(x_1, y_1, f(x_1, y_1))$, we compute:

$$f(x, y_0) \approx \frac{f(x_0, y_0)(x - x_0) + f(x_1, y_0)(x_1 - x)}{x_1 - x_0}$$

$$f(x, y_1) \approx \frac{f(x_0, y_1)(x - x_0) + f(x_1, y_1)(x_1 - x)}{x_1 - x_0}$$

Then the final step is:

$$f(x, y) \approx \frac{f(x, y_0)(y - y_0) + f(x, y_1)(y_1 - y)}{y_1 - y_0}$$

A similar, albeit far more ugly, derivation shows that the error of bilinear interpolation is also second-order. It is also possible to do trilinear interpolation by taking eight lookup table entries, reducing them to four values by interpolating in the x direction, reducing those to two points by interpolating in y, and performing a final interpolation in z. This scales up to n dimensions with 2^n lookup table entries.

Higher-order interpolation using splines (see Section 11.2) or polynomials is another technique in this toolbox that helps to improve accuracy, but is more computationally expensive than linear interpolation. Polynomial interpolation is not automatically continuous at the breakpoints between segments, but spline interpolation is easy to make continuous at these points. These methods trade off speed for accuracy or speed for table size.

A combination of lookup tables plus interpolation makes for an alluring method for function approximation, but can be surprisingly slow when functions can be approximated by other methods. The major advantage of lookup table approximations is that they allow you to approximate a function without any closed-form representation. Finally, lookup tables can give fast, accurate results when the hardware is available to support them.

10.5   Comparison of Math Libraries

Most math libraries do not provide correctly rounded approximations of transcendental functions, but get close enough to be useful. Due to dynamic linking, this can mean that code does not reproduce when moved between computers. If one computer's math library has a different accuracy level than the one on which a piece of software was tested and developed, that can cause "floating point error", even though the error is unrelated to the number format or the basic operations themselves. If you care about or need numerical accuracy, Statically linking your own selection of math library can help to avoid any headaches about this.

Compilers, chip vendors, and operating systems provide a default math library. These libraries are often designed to balance performance and accuracy, frequently eschewing "correctly rounded" for "good enough". The main exception to this philosophy is the LLVM math library, which favors correctly rounded functions. A few of the common math libraries are listed here, with a summary of their accuracy at the time of writing [20]:

- **GNU-libc math** contains a default math library for GCC, and has error under 4 ULPs for most functions.

- **LLVM libc math** is the default library for LLVM, and uses correctly rounded functions. LLVM's math functions are consistent across platforms when compiled.

- **Intel's IML** is very close to correctly rounded, with worst-case error close to 0.5 ULP for most functions.

- **Apple Libm** is faithfully rounded, providing error under 1 ULP.

- **Microsoft UCRT math** has comparable accuracy to GNU libm, with some functions having error bounds of 4 ULPs.

- **CUDA Libm**, Nvidia's library for CUDA, is slightly worse in worst-case accuracy than GNU Libm.

- **ROCm Libm**, for AMD GPUs, is comparable to CUDA Libm in accuracy.

The variability between these default libraries should indicate that if you would like consistent results, it is a good idea to statically link one of them or to use a custom math library.

There are several math libraries that are specifically made for gaming applications. These math libraries do not differentiate themselves in accuracy, but do provide more features than others. Many of these libraries will contain extensive libraries for geometry functions like shape intersections and quaternion rotations in addition to elementary functions. These math libraries frequently have SIMD as a first-class citizen, and only use single-precision floats. Accuracy of these libraries is usually worse than the generic math libraries.

- **DirectXMath** from Microsoft is a SIMD-focused library that is intended to be used for graphics applications and games. DirectXMath provides a number of fast estimators (third and fourth order approximations) for functions like the trigonometric functions as well as the "accurate" versions ($\approx$10th order approximations).

- **GLM** is the math library that ships with OpenGL and is intended for SIMD and graphics applications. Its accuracy and implementations are similar to DirectXMath.

Finally, several math libraries provide correctly rounded functions, but are sometimes incomplete and usually are slower than the default libraries. Some of these libraries have fast and slow paths, which can cause certain inputs to take significantly longer than other inputs. However, these libraries usually come with proofs of correctness and proofs of accuracy for functions. The major correctly rounded math libraries are:

- **CRLibm** is a now-deprecated library (as of 2015) of correctly rounded functions from the INRIA research institution. The functions in CRLibm come with corresponding computer proofs in Gappa of their accuracy.

- **CORE-MATH** is a project from INRIA that is a successor to the CRLibm project. The CORE-MATH project uses tools like Sollya to construct correctly rounded functions and corresponding proofs of correctness and accuracy.

- **RLibm** is a project from Rutgers University that provides alternative correctly rounded implementations of transcendental functions. Unlike CORE-MATH and CRLibm, the functions in RLibm are constructed using linear programming solvers, leading to very different solutions.

- **GNU MPFR** is a correctly rounded library from GNU for a type arbitrary-precision arithmetic called multiple-precision floating point. MPFR is not intended for default floating point formats, but its implementations have been adapted.

The use of correctly rounded functions often carries a performance hit, although CORE-MATH and LLVM have been able to produce implementations that are competitive in benchmarks with less-accurate math libraries. Still, the variance in performance caused by having fast and slow paths can be significant when determinism is important.

Math library developers have generally used the techniques discussed in this chapter (and offshoots of those techniques) to develop these libraries. Some will use extended precision in certain calculations, but in most cases, quantization error causes problems before you get to a correctly rounded implementation unless that is an explicit goal. Correctly rounded functions and new techniques are still active areas of research, so libraries will likely improve over time, albeit slowly.

10.6 Floating Point Bit Hacks

Finally, some of the most fun and interesting approximations are rather bad approximations, but they take advantage of bit hacks that are available with the floating point number system. A piece of code made famous by the source code of *Quake III* has captured the imagination of systems programmers and future numerical analysts alike [42]. A pseudocode version of this cryptic algorithm can be found in Algorithm 10.1.

Algorithm 10.1 *Quake III* Bit-hacking Reciprocal Square Root

Input: $x > 0$ in single-precision float
Output: $y = \frac{1}{\sqrt{x}}$ to 2–3 decimal places

1: $i \leftarrow \texttt{bitCast}_{\texttt{I32}}(x)$ ▷ Get bits of number to `int32`
2: $i \leftarrow \texttt{0x5f3759df} - (i \gg 1)$ ▷ "Magic" bit hacking
3: $y \leftarrow \texttt{bitCast}_{\texttt{F}}(i)$ ▷ Return to floating point, 3.5% accuracy

4: $x_2 \leftarrow x \times 0.5_\texttt{F}$ ▷ Set up for Newton-Raphson
5: $y \leftarrow y \times (1.5_\texttt{F} - x_2 \times y \times y)$ ▷ Refine, 3.5% → 0.17% accurate
6: **return** y

This is actually two numerical algorithms in one. The second algorithm, at line 4, is easier to comprehend than the first: It is a Newton-Raphson iteration. In this case, the refinement iteration, derived from the equation $\frac{1}{y^2} - x = 0$, is:

$$y_{i+1} = y_i - \frac{\frac{1}{y^2} - x}{\frac{-2}{y^3}} = y_i + \frac{y_i - xy_i^3}{2} = y_i \left(\tfrac{3}{2} - \tfrac{1}{2}xy_i^2\right)$$

The first section of the algorithm is the tough part. We start by bit casting the floating point number into integer space. We are not converting, just taking the bits of the floating point number and working with them as though they are an integer. If you recall Figure 2.5, this is effectively taking a logarithm of the floating point number, with an offset and some error. Right shifting by 1 divides this integer by 2, which has an interesting property in logarithmic space. With our integer $i \approx \log_2(x)$:

$$i \gg 1 \approx \frac{\log_2(x)}{2} = \log_2\left(\sqrt{x}\right)$$

Negating the log should give us:

$$-(i \gg 1) \approx -\log_2\left(\sqrt{x}\right) = \log_2\left(\frac{1}{\sqrt{x}}\right)$$

That seems good. When we move back into floating point space, we will be undoing the log, so we seem to have constructed $\frac{1}{\sqrt{x}}$. However, floating point exponents are not represented as signed integers. They are represented as unsigned integers with a bias. By moving into log space by bit casting to an integer, we are also operating on our bias. The right shift by 1 takes the square root of *everything* about x, including the bias. We started with a bias of 2^{127}, so now we have a number biased by $\sqrt{2^{127}}$. That is where the magic number comes into play. Breaking down the magic number further into its single-precision floating point parts:

$$
\begin{array}{cccc}
 & \texttt{S} & \texttt{Exp} & \texttt{Mantissa} \\
\texttt{0x5f3759df} = & \texttt{0} & \texttt{10111110} & \texttt{01101110101100111011111} \\
 & + & 2^{63} & 1.432430148124\ldots
\end{array}
$$

$$
2^{63} \times 1.432430148124\ldots \approx 2^{63} \times \sqrt{2} = \sqrt{2^{127}}
$$

This is not a random magic number at all! This is almost equal to the square root of the exponent bias in single-precision float, which effectively provides for a clean subtraction of our original number in log space. In order to get back to a properly biased floating point number, we have to get back to a bias of 2^{127} by multiplying by $\sqrt{2^{127}}$, which is done through an addition in log space (approximated by an integer addition). The magic number is mostly just a bias correction. The error in the mantissa of this magic number serves as a correction factor. Moving to integer space does not take the logarithm of the mantissa, so the trailing bits of the magic number have been twiddled to balance the error of this process as much as possible.

The fabled constant from the code of *Quake III* is not actually the best constant for this operation, however. Researchers have conducted constant searches more recently, and have found varying constants that are better by one metric or another, but it does not appear that the original choice of constant was optimized to any specific objective function [43]. My assumption is that the original author (not known to this day, but there are a few suspects) started with $\sqrt{2^{127}}$ and fiddled with bits by hand to balance error in the approximation.

Modifying this structure with a different manipulation of the input and a different magic number can give approximations of x^a for any integer a. The general form, which applies to all widths of integer and floating point, is:

$$
x^a \approx \texttt{bitCast}_{\texttt{F/D}}\left[\texttt{bias} + (\texttt{bitCast}_{\texttt{I32/I64}}(x) * a)\right]
$$

This is a neat trick, but it is unlikely to be useful for many values of a, and isn't useful for $a = -\frac{1}{2}$ (the original algorithm) thanks to instructions like RSQRTSS and VRSQRT14SD, which both compute a better approximation and do it faster [3]. It's hard to beat the hardware.

This sort of approximation can also apply with a fixed base. Exponentiation can be a relatively slow operation, but treating numbers in approximate

---

Algorithm 10.2 Bit-hacking Fast Exponential

Input: double-precision x
Output: double-precision $y = e^x$ accurate to within 6.2%

  1: $x \leftarrow x \times$ 6.497320848556798e15$_D$ ▷ "Magic" constant multiplication
  2: $i \leftarrow \texttt{toInt64}(x)$ ▷ Convert to fixed point
  3: $i \leftarrow i + (1023 \ll 52)$ ▷ Add exponent bias
  4: $y \leftarrow \texttt{bitCast}_D(i)$ ▷ Take approximate 2^x by conversion
  5: **return** y

---

log space can get us there quickly. For example, $\exp(x)$ for any x can be approximated by Algorithm 10.2.

The magic constant here is equal to $[2^{52} \times \log_2(e)]$ (the mathematical constant e, not an exponent) as calculated in double-precision floating point. This gives us a number that, when converted to integer, is a fixed-point representation of $[x \times \log_2(e)]$ with 52 bits (exactly the size of the mantissa) behind the decimal place. In other words, we are making a fixed-point exponent that aligns with the exponent field of a double-precision float. From there, biasing appropriately and then converting back to floating point by bit casting will give us a correct exponent but a linearized mantissa compared to $\exp(x)$.

You can do somewhat better on most error metrics by tuning the constants in this formula. This algorithm with the most meaningful constants will almost always produce an overestimate because the bit cast back to floating point is linear in the mantissa, while the exponent of the significand has some curvature. As a result, this magic number approximation will overestimate by as much as 6.2%, but will almost never underestimate. A better selection of constants can balance the error here, reaching an error under 3.1%. A similar function for a fast logarithm in base two, again with untuned but meaningful constants, is in Algorithm 10.3. With tuned constants, error can drop to 2.7% or lower.

Algorithm 10.3 Bit-hacking Fast Log Base 2

Input: double-precision x
Output: double-precision $y = \log_2(x)$ accurate to within 5.4%

  1: $i \leftarrow \texttt{bitCast}_{I64}(x)$ ▷ Bit cast to fixed point
  2: $i \leftarrow i - (1023 \ll 52)$ ▷ Unbias to get our number in two's complement
  3: $y \leftarrow \texttt{toDouble}(i)$ ▷ Take approximate $\log_2(x)$ by conversion
  4: $y \leftarrow y/2^{52}$ ▷ Divide out the fractional bits
  5: **return** y

It should be possible to imagine how one might compute even a function like `pow` approximately using a similar method, taking an approximate log of one argument to get the magic constant for the exponential in this algorithm.

Bit hacks will not win any awards for accuracy, but can be blazing fast for transcendental functions that happen to be easier to compute in log space.

Check Your Understanding

Problem 10.1. An ancient Indian approximation for the `sinPi` function is given by [44]:

$$\texttt{sinPi}(x) \approx \frac{16x(1-x)}{5 - 4x(1-x)}$$

Comparing against the double-precision `sinPi` function: What is the error of this approximation using minimax error? What is it using RMS error? What is the accuracy of this approximation in bits?

Problem 10.2. Construct a fourth-order Remez approximation of $f(x) = \log_2(x)$ with x in the region $[1, 2]$, and construct a lookup table approximation with linear interpolation in the same region. How large of a lookup table do you need to equal the accuracy of the Remez approximation?

Problem 10.3. Either hacking together a solver, using a solver, or guessing and checking, tune the constant from the exponential algorithm at the end of Section 10.6 for x in $[-100, 100]$. First, find the constant with the best minimax error. Compute another constant for RMS error in the region. Finally, find a third constant that produces the best accuracy.

Problem 10.4. Using double precision floating point, construct a third-order Remez approximation for $f(x) = 1/\sqrt[3]{x}$ in the region $[1, 8]$, and a Newton-Raphson refinement step. Adapt your solution to single precision with a wrapper that converts the input from single precision to double precision and a final conversion of the output to single-precision floating point.

Try your solution on every single-precision floating-point number in the region. How many rounds of Newton-Raphson do you need to reach a correctly rounded single-precision inverse cube root?

Problem 10.5. Devise a double-precision floating-point bit hack and a Newton-Raphson refinement step for $f(x) = 1/\sqrt[3]{x}$. What is the maximum error of your bit hack approach compared to a double-precision computation in the range between 1 and 2? With one round and two rounds of Newton-Raphson, what is the error? How does the speed compare to the exact version?

Problem 10.6. A common function to find in AI and ML is the "Swish" function:

$$S(x) = \frac{x}{1 + \exp(-x)}$$

Using any of the methods found in this chapter, make an approximation for the Swish function, focusing on the domain $[-6, 6]$. Compare speed and accuracy to a version using single-precision floating point, noting that AI and ML commonly use floating point formats with 4–5 bits of precision.

11

Geometry in Floating Point

From collision to rendering, geometric calculations are the most common calculations to find in video games. Many of these calculations are built around the idea of continuous space. In floating point, however, space is not continuous. This can cause graphical artifacts and glitches, such as holes in the geometry of a solid world that a player can fall through. Shapes will also slightly deform as they move through space due to rounding and other quantization-related errors, which can be a source of unexpected behavior in a system.

However, the way floating-point numbers work helps you in many cases. When division by zero is acceptable, many edge cases can actually be handled without special-case logic at all.

11.1   Polygons and Transforms

The exact shape of objects in floating point depend on how that shape is specified. A box specified by its eight vertices will look slightly different than a rectangle specified by a center point, an orientation, and its dimensions. Geometric objects are usually specified as collections of triangles for rendering, but simulators may use many different representations as required for computational convenience. Storing a shape as a collection of polygons now means storing a list of floating-point coordinates that indicate the vertices of the polyhedron plus a list of indices that specify the vertices used in each face.

It is generally preferable to work with shapes that are closed manifolds, which allows them to behave well with algorithms and work within the floating-point number space. This means placing restrictions on the geometry of shapes: Avoiding internal faces, edges and vertices that don't line up with each other, and connections between closed shapes that are joined only at an edge or a vertex. Meshes that are not manifolds can behave weirdly in floating point. Figure 11.1 shows how this process happens around a T-junction in a polygon mesh. For this reason, T-junctions and other structures that create zero-sized holes are considered bad practice for 3D modelers. They create a zero-sized hole, breaking the manifold constraint, which in turn becomes a non-zero-sized hole when transformed. Figure 11.2 shows a repaired version

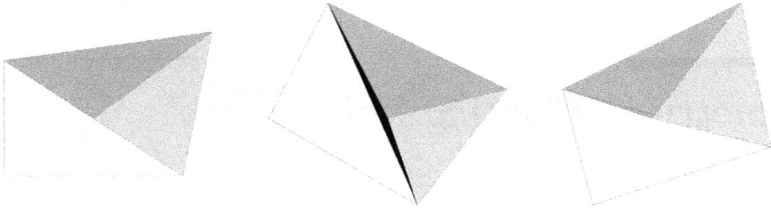

FIGURE 11.1
T-junction in floating-point geometry (left), which creates a hole when rotated in one orientation (center), and an overlap in another orientation (right) due to deformation from rounding.

of the T-junction. If shapes are modeled as manifolds, they will remain closed and well-behaved despite rounding and other numerical inaccuracies.

Transformations of shapes are usually specified as four-dimensional matrix multiplications with floating-point matrices and vectors. The use of a fourth pseudo-dimension allows arbitrary geometric transformations to be done with invertible matrices, as well as enabling calculations like perspective projection to be simple matrix algebra. The fourth dimension can be thought of as a "denominator" coordinate, where the vector (x, y, z, w) indicates the point at $\left(\frac{x}{w}, \frac{y}{w}, \frac{z}{w}\right)$. This coordinate also allows position vectors to be distinguished from direction vectors. Position vectors have $w = 1$, while direction vectors have $w = 0$, indicating a point at infinity in the direction of (x, y, z). Most transformations thus leave the fourth dimension untouched, only using it when needed. This four-dimensional system of coordinates is called **homogeneous coordinates**.

For example, a translation of a shape by a vector t followed by a rotation about the x-axis with angle r is done with the following pair of matrix

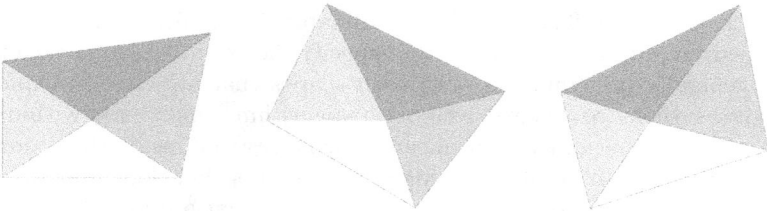

FIGURE 11.2
The mesh from Figure 11.1 repaired by adding an extra triangle. Under the same transformations, rounding still deforms the shape, but there are no holes or overlaps.

multiplications:

$$
\begin{bmatrix} x \\ y \\ z \\ w \end{bmatrix} = \begin{bmatrix} 1 & 0 & 0 & 0 \\ 0 & \cos(r) & -\sin(r) & 0 \\ 0 & \sin(r) & \cos(r) & 0 \\ 0 & 0 & 0 & 1 \end{bmatrix} \begin{bmatrix} 1 & 0 & 0 & t_x \\ 0 & 1 & 0 & t_y \\ 0 & 0 & 1 & t_z \\ 0 & 0 & 0 & 1 \end{bmatrix} \begin{bmatrix} x_0 \\ y_0 \\ z_0 \\ 1 \end{bmatrix}
$$

For reference, Table 11.1 contains a list of some common matrix multiplications used as geometric transformations.

Matrix multiplication is computed as a series of dot products of the rows of one matrix with the columns of the other:

$$
\begin{bmatrix} \underline{\mathbf{a}} \\ \underline{\mathbf{b}} \\ \underline{\mathbf{c}} \\ \mathbf{d} \end{bmatrix} \begin{bmatrix} \mathbf{x} & \mathbf{y} & \mathbf{z} & \mathbf{w} \end{bmatrix} = \begin{bmatrix} \mathbf{a \cdot x} & \mathbf{a \cdot y} & \mathbf{a \cdot z} & \mathbf{a \cdot w} \\ \mathbf{b \cdot x} & \mathbf{b \cdot y} & \mathbf{b \cdot z} & \mathbf{b \cdot w} \\ \mathbf{c \cdot x} & \mathbf{c \cdot y} & \mathbf{c \cdot z} & \mathbf{c \cdot w} \\ \mathbf{d \cdot x} & \mathbf{d \cdot y} & \mathbf{d \cdot z} & \mathbf{d \cdot w} \end{bmatrix}
$$

In turn, a dot product is a sequence of FMA operations that perform the sum of an element-wise product over the vector, shown here for 3-element vectors:

$$
\mathbf{a \cdot x} = \text{FMA}\left[a_3, x_3, \text{FMA}(a_2, x_2, a_1 \times x_1)\right]
$$

Each additional dimension adds an FMA to the matrix multiplication. Thus, a general four-by-four matrix multiplication comes down to 64 basic operations: 48 FMAs and 16 multiplications. This is simplified when you have large numbers of zeros, as in the matrices in the example above, but the order of operations can still matter, and there is nominally some rounding between each one. A matrix-vector product with four dimensions is a combination of only four dot products, using 16 total operations.

Matrix multiplication is a highly optimized operation in every computing platform, with special attention paid to four-by-four matrices for obvious reasons. Matrix layouts and exact sequences of FMA operations are heavily optimized around loading and storing data to maximize cache efficiency and make full use of vector-processing units. Graphics calculations are also often done with a form of fast math, where matrix algebra is allowed, effectively treating floating point as associative. A string of matrix operations may then be pre-computed into a single combined transform that is faster to apply. The final single matrix multiplication decomposes into products and sums, which are often computed either using FMA operations or dedicated matrix-multiply units that are available in GPUs [12]. These dedicated units can sometimes operate in mixed precision, accumulating their results in a wider format than the input, which prevents error.

TABLE 11.1
Common transformation matrices for three-dimensional objects in homogeneous coordinates. All assume coordinates are stored as column vectors, and should be transposed for row vectors. Transforms are combined by matrix multiplication, so an arbitrary rotation is the product of all three rotations.

Transformation	Matrix Form
Translation by (t_x, t_y, t_z)	$\begin{bmatrix} 1 & 0 & 0 & t_x \\ 0 & 1 & 0 & t_y \\ 0 & 0 & 1 & t_z \\ 0 & 0 & 0 & 1 \end{bmatrix}$
Scaling by (s_x, s_y, s_z)	$\begin{bmatrix} s_x & 0 & 0 & 0 \\ 0 & s_y & 0 & 0 \\ 0 & 0 & s_z & 0 \\ 0 & 0 & 0 & 1 \end{bmatrix}$
Rotation by r about the x-axis	$\begin{bmatrix} 1 & 0 & 0 & 0 \\ 0 & \cos(r) & -\sin(r) & 0 \\ 0 & \sin(r) & \cos(r) & 0 \\ 0 & 0 & 0 & 1 \end{bmatrix}$
Rotation by r about the y-axis	$\begin{bmatrix} \cos(r) & 0 & \sin(r) & 0 \\ 0 & 1 & 0 & 0 \\ -\sin(r) & 0 & \cos(r) & 0 \\ 0 & 0 & 0 & 1 \end{bmatrix}$
Rotation by r about the z-axis	$\begin{bmatrix} \cos(r) & -\sin(r) & 0 & 0 \\ \sin(r) & \cos(r) & 0 & 0 \\ 0 & 0 & 1 & 0 \\ 0 & 0 & 0 & 1 \end{bmatrix}$
Arbitrary shear transformation s_{ab} shears the a-axis by the b-axis	$\begin{bmatrix} 1 & s_{xy} & s_{xz} & 0 \\ s_{yx} & 1 & s_{yz} & 0 \\ s_{zx} & s_{zy} & 1 & 0 \\ 0 & 0 & 0 & 1 \end{bmatrix}$
Perspective projection	$\begin{bmatrix} 1 & 0 & 0 & 0 \\ 0 & 1 & 0 & 0 \\ 0 & 0 & -1 & 0 \\ 0 & 0 & -1 & 0 \end{bmatrix}$

11.2 Splines

An alternative to using polygons based on line segments is to use more parameters to define curves. This method allows smooth surfaces, which would ordinarily need a large number of polygons, to be defined with many fewer control points, as shown in Figure 11.3. An efficient way to do this is to use a family of geometric objects known as splines. Splines are sequences of second- or third-order curves that are defined using control points and a set of polynomial functions that dictate the weighting of each control point at any given point, so that they are continuous and differentiable across the surface, making smooth shapes. Second- and third-order curves are used since they balance visual "smoothness" and algebraic properties (such as differentiability for third-order curves) against computational difficulty to compute the position of the spline.

A spline is then drawn by interpolating between the control points, weighted by a set of polynomial functions. Multidimensional splines are created by having a manifold of control points rather than a sequence, and using multidimensional weight functions. There are multiple types of splines, where the type of spline dictates the family of weighting functions to use during interpolation. Since control points are used to define the shape of splines, splines can be deformed by applying the transformation matrices of Table 11.1 to the control points, and the spline will follow.

Games don't often use splines for three-dimensional geometry, but animated movies and CAD programs will use them to trade off computation speed for visual accuracy even when the view window is zoomed in. Two-dimensional graphics like fonts and GUIs will frequently use splines to define their geometry, allowing artists to create clean curves that render sharply at any resolution. Games will sometimes also use splines to define movement paths and represent objects that are "one-dimensional", such as strings or

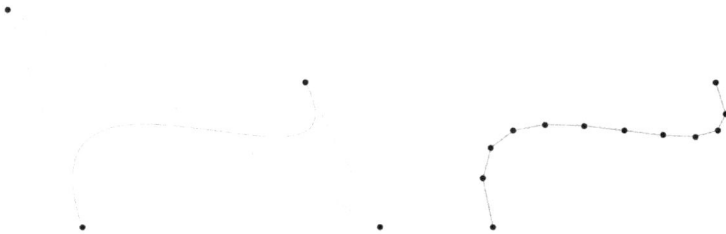

FIGURE 11.3
Comparison between a spline curve (left) and a tessellated mesh (right) following the same curve. The spline has four control points and is smoother than the tessellated version with 12 vertices.

ropes. Vector drawing systems have **Bézier curves** as a first-tier primitive, which are the go-to spline for efficient calculation (Figure 11.3 has a Bézier curve). CAD programs will often use a more flexible form of spline surface called a **NURBS** (Non-Uniform Rational B-Spline) surface, which gives the user a lot of control to define all types of curves that might be present in a mechanical part. Both of these fall into the family of splines, but have different weighting functions for their control points.

By turning fine geometry into coarse-grained curves, splines allow you to avoid floating-point deformation in objects that need to be represented accurately. Small deviations in control point locations have less weight on the shape of objects than deviations in polygon vertices, and the algorithms for drawing and displaying splines stay accurate with some jitter in operations. However, this comes at a cost of computing power.

11.3　Bézier Curves and Surfaces

Bézier curves are splines based on the Bernstein polynomials. A Bézier curve of order n has $(n + 1)$ control points, and uses these control points and the order-n Bernstein polynomials to define the shape of the curve. The Bernstein polynomials of order n have the form:

$$b_{v,n}(t) = \binom{n}{v} t^v (1 - t)^{n-v} \qquad\qquad v = 0, 1, \ldots, n$$

When $v = 0$, the Bernstein polynomial is weighted toward $t = 0$ and goes to zero at $t = 1$. As v increases, the vth Bernstein polynomial of order n peaks later and later as the $(1 - t)$ factor gets weaker and the t factor gets stronger, with the polynomial with $v = n$ peaking at $t = 1$. The third-order Bernstein polynomials are shown in Figure 11.4, which are the polynomials used to define a third-order Bézier curve (the most common Bézier curve). Like the Chebyshev polynomials, the Bernstein polynomials of degree n are an orthogonal basis for all polynomials of degree n, but they operate in the window $0 \leq t \leq 1$.

Described using the Bernstein polynomials, a Bézier curve is defined by the sum:

$$\mathbf{x}(t) = \sum_{i=0}^{n} \mathbf{c}_i b_{i,n}(t) \qquad\qquad 0 \leq t \leq 1$$

where $\mathbf{c}_i$ is the ith control point for the curve, and b is the Bernstein polynomial from above. The Bézier curve thus linearly interpolates a weighted average of the control points, weighted by the Bernstein polynomials. A Bézier surface, a smoothly curved sheet in three-dimensional space, can be built from

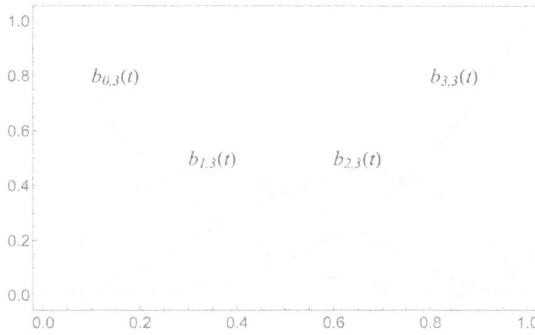

FIGURE 11.4
The third-order Bernstein polynomials, which form the basis for the most common form of Bézier curves.

the one-dimensional curve equation by adding a second dimension to the parameter space and the space of control points (a tensor product of two Bézier curves):

$$\mathbf{x}(t,u) = \sum_{i=0}^{n} \sum_{j=0}^{n} \mathbf{c}_{i,j} b_{i,n}(t) b_{j,n}(u) \qquad 0 \leq t, u \leq 1$$

A Bézier surface of order n uses an $n \times n$ grid of control points, and the product of two Bernstein polynomials in the two parametric dimensions to weight the control points.

If a shape needs to be more complicated than a single Bézier curve of order n can define, multiple Bézier curves can be used to draw it, and can join smoothly as shown in Figure 11.5, by setting the first two control points of the joined Bézier curve to continue the curve and keep its first derivative going. There may be a discontinuity in the second derivative at the point of the join. A higher-order Bézier curve is also an option, but higher-order curves take more computing power to draw, so it is more efficient to use a sequence of third-order curves when you need a continuous and smooth shape than to use a higher-order Bézier curve. Note that it is impossible to draw a perfect circle or ellipse with a Bézier curve of any order, but a set of a few Bézier curves can approximate these shapes very closely [45].

Mirroring the Chebyshev approximation, which is not computed using the Chebyshev polynomials, Bézier curves are not drawn using the Berenstein polynomials. The utility of the Bézier curves is that there is a fast algorithm for rendering them, which is built on recursive linear interpolation. This algorithm is named for its inventor, French physicist Paul De Casteljau. **De Casteljau's algorithm** draws a Bézier curve by interpolating between the control points to find a new set of $n-1$ points, and then interpolating between these new control points to find $n-2$ points, and repeating until there is only

FIGURE 11.5
A pair of joined Bézier curves. The second Bézier curve (control points in gray) shares a control point with the first curve (black control points) and its second control point forms a line with the last control points of the first curve.

one point. De Casteljau's algorithm is described for third-order Bézier curves in Algorithm 11.1, and a visual example is shown in Figure 11.6. This algorithm has no direct polynomial evaluations, but a look at the math of linear interpolation will show that it is equivalent [45].

Algorithm 11.1 De Casteljau's Algorithm for Third-order Bézier Curves

---

Input: t where $0 \leq t \leq 1$, a set of four control points $\mathbf{c}_0, \ldots, \mathbf{c}_3$
Output: point $\mathbf{x}$ at position t along the Bézier curve defined by $\mathbf{c}_0, \ldots, \mathbf{c}_3$

1: $\mathbf{a}_0 \leftarrow \mathtt{lerp}(\mathbf{c}_0, \mathbf{c}_1, t)$ ▷ First set of interpolations
2: $\mathbf{a}_1 \leftarrow \mathtt{lerp}(\mathbf{c}_1, \mathbf{c}_2, t)$
3: $\mathbf{a}_2 \leftarrow \mathtt{lerp}(\mathbf{c}_2, \mathbf{c}_3, t)$
4: $\mathbf{b}_0 \leftarrow \mathtt{lerp}(\mathbf{a}_0, \mathbf{a}_1, t)$ ▷ Second set of interpolations
5: $\mathbf{b}_1 \leftarrow \mathtt{lerp}(\mathbf{a}_1, \mathbf{a}_2, t)$
6: $\mathbf{x} \leftarrow \mathtt{lerp}(\mathbf{b}_0, \mathbf{b}_1, t)$ ▷ Final interpolation
7: **return x**

While Algorithm 11.1 is relatively short, each linear interpolation takes two FMA operations in each dimension, so a two-dimensional Bézier curve needs 24 FMA operations to find each point along the line. The link between

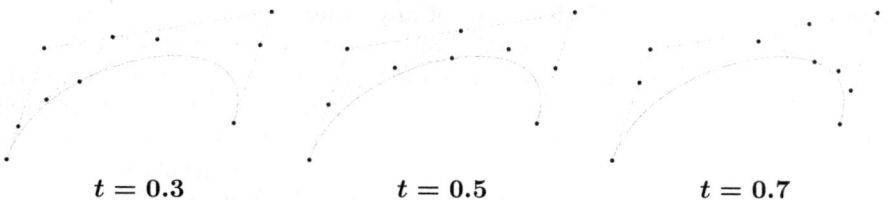

$t = 0.3$ $t = 0.5$ $t = 0.7$

FIGURE 11.6
Finding three points along a Bézier curve using De Casteljau's algorithm.

a change in t and movement in space is also not linear, so programs that render Bézier curves or surfaces must account for the fact that they need to move at a non-constant velocity. However, Bézier curves stay within the convex hull of their control points, so calculations such as collision detection can operate on the control points rather than the curve itself. Generalizing Algorithm 11.1 is a matter of adding or removing interpolation rounds to fit the order of the curve (or looping for variable order), so drawing a Bézier curve of order n is an $O(n^2)$ operation.

11.4   B-Splines and NURBS

Bézier curves use the Bernstein polynomials, which are nonzero throughout the entire range of the curve. This means that increasing the number of control points in one spline causes an increased computation cost, and complex shapes have to be constructed out of multiple Bézier curves. The family of B-splines address this problem by using windowed polynomials that decay to zero outside of a short range. This allows a single B-spline to represent a shape of high complexity by adding more control points to one spline.

A B-spline starts with a vector of values of t called knots, which are the values of the parameter t where the polynomial segments of the spline are stitched together. Like Bézier curves, t usually ranges from zero to one, but B-splines do not require any specific range. In computer systems, it may be better to use integer knot positions so that they are represented exactly. The knot vector of a B-spline is the sorted collection of all of the knots:

$$T = \{t_0, t_1, t_2, \ldots, t_m\}$$

A B-spline with degree n can have more than m knots. The typical form of a B-spline has $(m + n + 1)$ knots spaced equidistantly across the range of t with $(n + 1)$ stacked knots at the endpoints of the spline covering the boundary conditions. This is called a **cardinal B-spline**. An example of a cardinal B-spline, showing its control points and knots, is shown in Figure 11.7.

The B-splines use a basis function that is defined recursively in the order of the spline, and is based on the position of the knots:

$$B_{i,0}(t) = \begin{cases} 1 & t_i \leq t < t_{i+1} \\ 0 & \text{otherwise} \end{cases}$$

$$B_{i,n}(t) = \frac{t - t_i}{t_{i+n} - t_i} B_{i,n-1}(t) + \frac{t_{i+n+1} - t}{t_{i+n+1} - t_{i+1}} B_{i+1,n-1}(t)$$

The basis for the zeroth-order B-splines is a function that is one in the corresponding segment of the spline and zero everywhere else. This makes a zeroth-order B-spline equivalent to a piecewise function in t, and a first-order

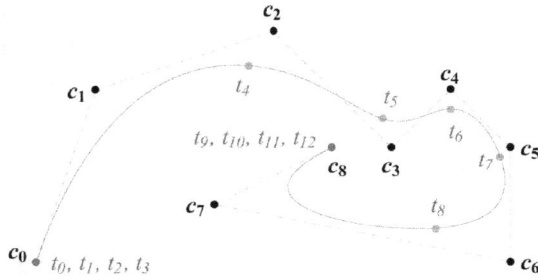

FIGURE 11.7
A cardinal B-spline of order 3 with 9 control points and 13 knots. The knots and control points are labeled, with knots in gray and control points in black.

B-spline a polygon of the control points. The higher-order B-splines then linearly interpolate between the lower-order knots, as shown in Figure 11.8. Like Bézier curves, most uses of B-splines opt for a third-order spline. The basis functions Cardinal B-splines are shown in Figure 11.9.

Finally, the spline itself is constructed as the weighted average of its control points, weighted by the B functions that vary with our parameter, t:

$$\mathbf{x}(t) = \sum_{i=0}^{m} \mathbf{c}_i B_{i,n}(t)$$

The windowing caused by the definition of the B polynomials means that the evaluation of this sum is now $O(nm)$, since each possible value of t has only n nonzero B functions. The evaluation is no longer quadratic in the number of control points.

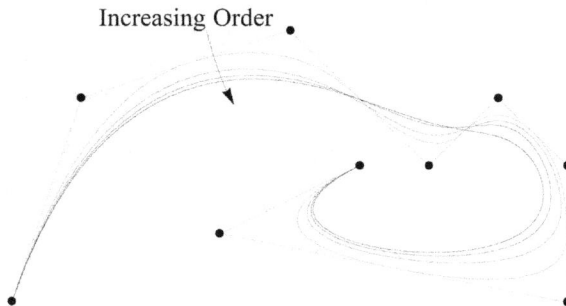

FIGURE 11.8
Cardinal B-splines sharing the same control points (black points) going from order 1 to order 5 (light to dark). The line gets more curved and individual control points are less prominent for higher-order curves.

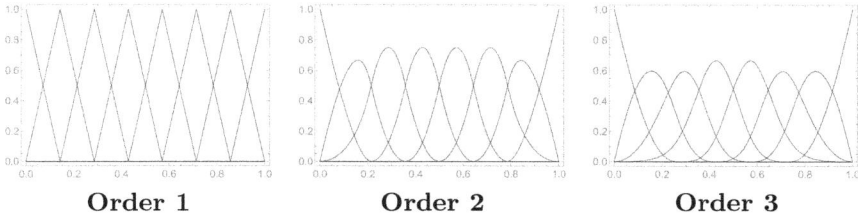

Order 1   **Order 2**   **Order 3**

FIGURE 11.9
Basis functions used for cardinal B-splines of orders 1–3 with 8 control points.
The sum of all basis functions is 1 for all values of t.

A non-uniform rational B-spline (NURBS) generalizes further, and adds
a weight to each control point (making the spline non-uniform). Since the
weighted B functions are no longer guaranteed to sum to one, we then have
to divide by the sum of the weights to normalize (making them rational):

$$\mathbf{x}(t) = \frac{\sum\limits_{i=0}^{m} \mathbf{c}_i w_i B_{i,n}(t)}{\sum\limits_{i=0}^{m} w_i B_{i,n}(t)}$$

Like a Bézier curve, this generalizes to multi-dimensional surfaces by taking
a cross product of NURBS splines [46], resulting in surfaces like Figure 11.10:

$$\mathbf{x}(t, u) = \frac{\sum\limits_{i=0}^{k} \sum\limits_{j=0}^{l} \mathbf{c}_{i,j} w_{i,j} B_{i,n}(t), B_{j,n}(u)}{\sum\limits_{i=0}^{k} \sum\limits_{j=0}^{l} w_{i,j} B_{i,n}(t), B_{j,n}(u)}$$

There is an equivalent algorithm to De Casteljau's algorithm for quickly
drawing NURBS splines and surfaces. De Boor's algorithm is this extension,
and applies to general B-splines. Extending De Boor's algorithm to NURBS
is relatively simple when using homogeneous coordinates. This algorithm, in
an optimized version, is given in Algorithm 11.2. In this algorithm, lines 1–5
find the relevant set of control points and knots, and store weighted control
points as homogeneous coordinates. Lines 6–11 of this algorithm are similar
to Algorithm 11.1 in function, although we have changed the interpolation to
fit the basis functions for a B-spline and generalized to n dimensions. Finally,
we re-normalize away from homogeneous coordinates to get the final result.
Note that if we don't use homogeneous coordinates, we reduce to drawing a
B-spline (all $w_i = 1$).

Due to their flexibility and the ability to create arbitrary shapes with a
single spline surface, NURBS is the default form of spline used for rendering
and graphics. The use of weighting means that geometric shapes like circles

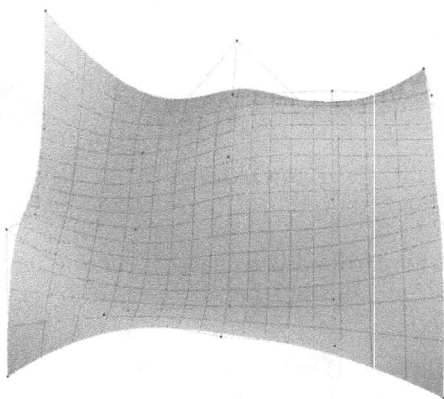

FIGURE 11.10
Example NURBS surface from a control point mesh. Each horizontal and vertical slice of the surface is a NURBS curve.

are much easier to represent with NURBS, and also that a NURBS surface can have both sharp corners and smooth curves, depending on the weight assigned to each control point. This allows generic geometry to be represented with a minimum number of weighted control points on a single spline.

The arrangement of arithmetic operations in De Casteljau's algorithm and De Boor's algorithm allows for the additions in each linear interpolation step to have orders-of-magnitude similar to the control points. Objects that are small and near the origin will usually be more precisely drawn, but the error will scale with the size of the curve and distance from the origin.

11.5 Faraway Objects and Floating Point

Recalling Figure 9.2, geometry loses resolution as we move away from the origin of our coordinate system. This also happens within camera calculations, where the origin of the calculation effectively moves to the position of the camera. As geometry gets further from the origin, it naturally loses resolution. This shows up in graphics pipelines in several places, and we will be exploring its effects on simulations further in Chapter 12.

This effect can also deform geometry, like the rotations of Figure 11.2, and affect how shapes behave under other transformations. Objects that are small and far from the origin do not benefit from the full precision of the number system. Further, operations that lose ULPs of precision can cause

---

Algorithm 11.2 De Boor's Algorithm for NURBS Curves

Input: t where $0 \leq t \leq 1$, a NURBS curve of order n defined by a set of control points $\{\mathbf{c}_0, \ldots, \mathbf{c}_m\}$, weights $\{w_0, \ldots, w_m\}$, and knots $T = \{t_0, \ldots, t_m\}$

Output: point $\mathbf{x}$ at position t along the NURBS curve

1: $k \leftarrow \texttt{findIndex}(t, T)$ $\triangleright$ Binary search T to find span containing t
2: Make set $D = \{\mathbf{d}_0, \ldots, \mathbf{d}_n\}$ $\triangleright$ Holds working set of control points
3: **for** j from 0 up to n **do** $\triangleright$ Gather in homogeneous coordinates
4:      $\mathbf{d}_j \leftarrow \texttt{concat}(\mathbf{c}_{j+k-n} \times w_{j+k-n}, w_{j+k-n})$
5: **end for**
6: **for** r from 1 up to n **do** $\triangleright$ n rounds of linear interpolation
7:      **for** j from n down to $(r-1)$ **do** $\triangleright$ Interpolate between knots
8:         $z \leftarrow (t - t_{j+k-n})/(t_{j+k-r+1} - t_{j+k-n})$
9:         $\mathbf{d}_j \leftarrow \texttt{lerp}(\mathbf{d}_{j-1}, \mathbf{d}_j, z)$
10:     **end for**
11: **end for**
12: $\{\mathbf{x}, w_x\} \leftarrow \mathbf{d}_n$ $\triangleright$ Our final result is at the nth $\mathbf{d}$
13: **return** $\mathbf{x}/w_x$ $\triangleright$ Renormalize from homogeneous coordinates

---

greater visual distortion on objects with lower resolution, as the operations round further from the exact answer.

Several steps in a rendering pipeline, such as z-buffering and z-culling, rely on precisely knowing the distance between the camera and the objects in the scene [47]. Recalling Table 11.1, the third row of the perspective projection matrix calculates and saves a signed distance from the camera along the negative z-axis. However, this row is usually scaled so that nearby objects that are still far enough from the camera to render have a value of $z = 0$, while faraway objects approach $z = 1$. That scale factor comes in using a near view plane setting the minimum draw distance and the far view plane setting the maximum draw distance. Using f as the distance to the far plane and n as the distance to the near plane along the negative z axis, a common z-axis scaling function looks like:

$$\hat{z} = \frac{f}{f-n}\left(\frac{n}{z}+1\right)$$

This function maps the z range to $\hat{z}$ so that $z = -f$ gives $\hat{z} = 1$, and $z = -n$ gives $\hat{z} = 0$, and operates quickly as a matrix multiplication in homogeneous coordinates. However, this scaling compresses the range of z values and also operates non-linearly in z, both of which introduce aliasing in $\hat{z}$ when considered in floating point. A graph of this function and a quantized version is shown in Figure 11.11. Faraway objects, which have the widest z range, will be compressed into a narrow range of $\hat{z}$ values that are close to one. That introduces an effect called z-fighting, which can create graphical glitches as shown in Figure 11.12.

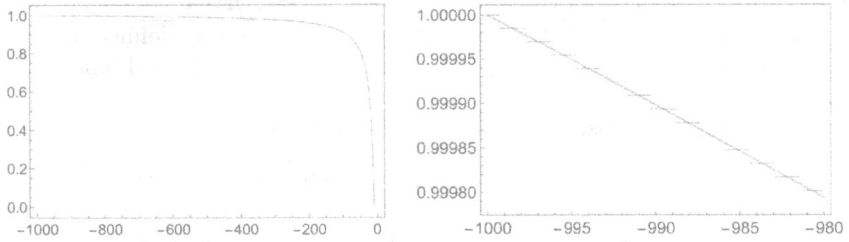

FIGURE 11.11
Plot of the z-axis mapping function used to normalize z values in perspective projections. Left: The mapping function over its entire range with $f = 1\,000$ and $n = 10$. Right: A zoomed-in version of the range of $z < -980$, with the exact function in gray and a 16-bit quantized version in black, showing how z-fighting arises.

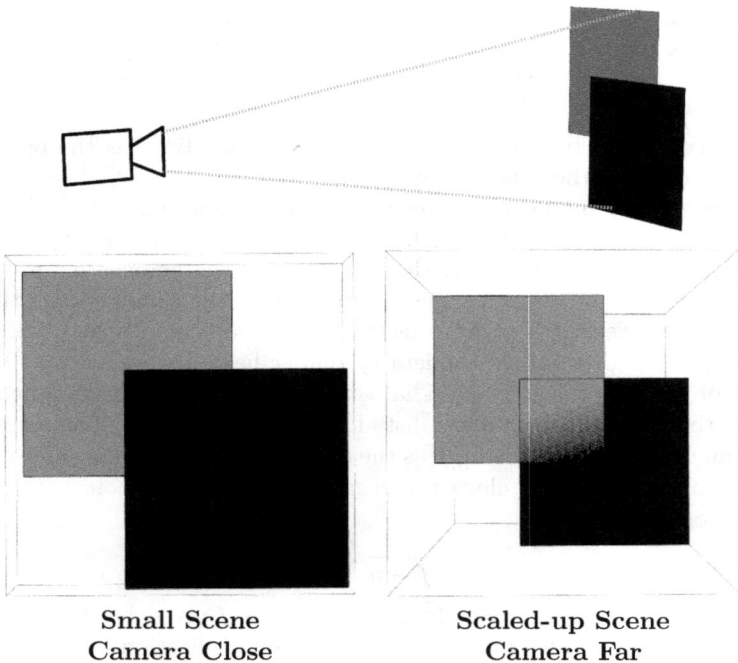

Small Scene
Camera Close

Scaled-up Scene
Camera Far

FIGURE 11.12
Top: Side view of a scene with a camera looking at two planes that are at different z values. Bottom Left: Camera views when close to the planes, with no z-fighting. Bottom Right: View when the scene is scaled up so that the camera is far away, where the overlap between the planes exhibits a graphical glitch from z-fighting.

A very simple solution to this problem is to use a linear mapping for $\hat{z}$:

$$\hat{z} = \frac{z - n}{f - n}$$

However, the advantage of the previous solution is that it can be done using the perspective projection matrix and a matrix multiply, while the new calculation for $\hat{z}$ needs to use a separate calculation. This also does not solve the issue of aliasing due to range compression: The z range of interest is still between n and f, while $\hat{z}$ still ranges between 0 and 1. While the $\hat{z}$ range probably has more numbers in it, they are concentrated toward zero and many will be inaccessible from the linear map.

A better solution to this is to use log-scaling in the normalization, which matches the scaling function to the density of the floating point numbers, minimizing the chance of collision. Most importantly, it expands the range of z values from $\hat{z} \in [0, 1]$ to $\log \hat{z} \in [0, -\infty]$ (a closed interval since infinity is a number), with large numbers for nearby objects, and values approaching zero as z grows:

$$\log \hat{z} = a \log(bz + 1)$$

where a and b are constants related to the positions of the view planes. This technique is present in several game engines, with an early example coming from *Just Cause 2*, an open-world game with a long rendering distance [48]. The use of the log scale is more expensive than the use of the linear scaling, but it mitigates issues from z-fighting because there are at least as many floating-point numbers in the range of $\log \hat{z}$ as there are in the range of the negative z axis. Log scaling does not eliminate z-fighting, since certain values of a and b can still cause two values of z to map to one value of $\hat{z}$, but it does alleviate the problem.

11.6 Intersection Finding and Collision Detection

Finding the intersections between shapes is a basic function involved in collision detection, which in turn is one of the major consumers of computing power from games to molecular dynamics simulations. To save computing power, collision detection usually uses simple shapes, like spheres or bounding boxes that surround more complex geometry. These shapes are also usually convex, which simplifies the mathematics of collisions significantly. Collision detection then reduces to finding the times and places when these bounding shapes intersect [49].

This can then be done by stopping at discrete points in time (frames) and checking collisions statically by finding where the edges of one shape intersect with the other shape, or it can be done by calculating the movement path of the vertices of each shape relative to the rest of the objects in the scene and finding the time when that movement path intersects with other objects. The latter approach is more expensive, but more accurate, and prevents artifacts like objects "phasing" through each other if they would only collide at a point between two frames.

In both cases, the problem of collision detection reduces to finding the intersection of a ray with an object. The second object is usually either a sphere or a set of planes. In the case of a ray and a plane, we define the ray as:

$$\mathbf{r}(t) = \mathbf{p} + t\mathbf{v} \qquad\qquad 0 \le t$$

One of the simplest shapes in collision detection is an axis-aligned bounding box (abbreviated AABB). The boundaries of the box are defined by six numbers: x_{min}, x_{max}, y_{min}, y_{max}, z_{min}, and z_{max}. Finding a collision is as simple as solving for the values of t where the ray reaches each of these bounds, as shown in Figure 11.13. There is a collision if there exists a value of t for which all dimensions of $\mathbf{r}$ are within the bounds that specify the bounding box. In equation form, for one dimension, the intersections with both bounds are:

$$t_{x_{min}} = \frac{x_{min} - p_x}{v_x} \qquad\qquad t_{x_{max}} = \frac{x_{max} - p_x}{v_x}$$

There is no problem if $v_x = 0$ here. If the x velocity of the shape is zero, we will divide by zero and properly indicate that the t value of the collision is infinity. However, if $(x_{min} - p_x)$ is also zero, indicating that the initial point of the ray is right on the boundary of the bounding box and is parallel to the x-axis, we will have a case of $\frac{0}{0}$, which will give us NaN. Thus, NaN is guaranteed to indicate an intersection with the plane, while an infinity will indicate no intersection—we would like edges of bounding boxes to indicate collision to close gaps that might exist between two adjacent bounding boxes. If we ensure that the $\mathbf{v}$ vector and the bounds of the bounding boxes cannot be negative zero, then we will also never produce negative infinity.

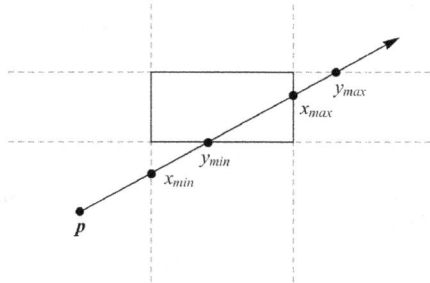

FIGURE 11.13

A ray passing through an axis-aligned bounding box. Any collision occurs when all dimensions of the ray are between the bounds of the bounding box.

To complete our intersection check, we need to sort the intersection times. If we enter the x, y, and z bounds of the bounding box before leaving any of them, we will indicate a collision. Otherwise, we indicate no collision. We need to sort as though the ordering looks like:

$$\texttt{NaN} < n < \infty$$

This can be done by sorting using the `minimum` function to find the minimum of numbers and the `maximumNumber` function to find the maximum. We can complete this calculation without branching for any numerical special cases. Tangent lines and other weird edge cases are handled by the number system.

Expanding to generic bounding boxes, a bounding box is usually defined by six planes, organized in pairs. This is similar to the axis-aligned case, but our intersection between the ray and the plane is now more difficult. Given a point **a** that is on the plane and a unit normal vector that is perpendicular to the plane, **n**, the equation for finding the intersection of a plane finds the time where the vector between **a** and $\mathbf{r}(t)$ is perpendicular to **n**. That guarantees that $\mathbf{r}(t)$ is on the plane as shown in Figure 11.14. We can find that point with a dot product:

$$[\mathbf{r}(t) - \mathbf{a}] \cdot \mathbf{n} = 0$$

Rearranging and substituting for $\mathbf{r}(t)$, we find the t value by:

$$t = \frac{(\mathbf{p} - \mathbf{a}) \cdot \mathbf{n}}{\mathbf{v} \cdot \mathbf{n}}$$

We get the same edge cases here as in the axis-aligned case. In fact, this equation is a generalized form of the intersection with an axis-aligned plane. It has the same edge cases: We will get a `NaN` if we have **v** going along the plane and **p** sitting on the plane, and we will get infinity if **v** is parallel to

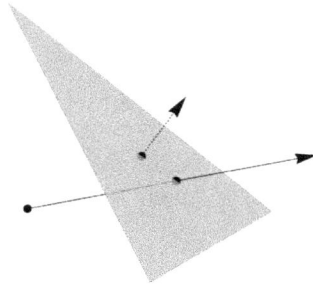

FIGURE 11.14
A ray passing through a triangular section of a plane in 3D space. The normal vector for the plane is the dashed arrow, and the point **a** is at its base. Note that the normal vector to the plane is perpendicular to the vector from **a** to the intersection point.

the plane and $\mathbf{p}$ is not on the plane. Using the sorting method above, we can compute collisions for arbitrary bounding boxes without any branching by allowing floating point to handle the (in this case, literal) edge cases.

The same goes for collision with a sphere. A sphere is defined by a center point, $\mathbf{c}$, and a radius, and we are looking for the value of t that minimizes the distance between $\mathbf{r}(t)$ and the center point. If that distance is less than or equal to the radius, we have a collision. The distance is:

$$d(\mathbf{c}, \mathbf{r}(\mathbf{t})) = \sqrt{(\mathbf{p} + \mathbf{v}t - \mathbf{c})^2}$$

We find the point of minimum distance by taking the derivative in t and setting equal to zero:

$$d'(\mathbf{c}, \mathbf{r}(\mathbf{t})) = \frac{\mathbf{v} \cdot (\mathbf{v}t + \mathbf{p} - \mathbf{c})}{\sqrt{(\mathbf{p} + \mathbf{v}t - \mathbf{c})^2}}$$

d' will never go to zero based on the denominator, so we get a zero when:

$$0 = \mathbf{v} \cdot (\mathbf{v}t + \mathbf{p} - \mathbf{c}) = |\mathbf{v}|^2 t + \mathbf{v} \cdot (\mathbf{p} - \mathbf{c})$$

Thus, the value of t that gets closest to the center of the sphere is:

$$t = \frac{\mathbf{v} \cdot (\mathbf{c} - \mathbf{p})}{|\mathbf{v}|^2}$$

If $\mathbf{v}$ is a unit vector, there is no division here at all. We would normally assume that a ray has nonzero velocity, but if it does, this calculation will yield a NaN, which will allow us to safely assume that the point of closest interaction is at $t = 0$. In most architectures, this check and replacement can be done without branching by using AND masks or conditional instructions. Note that there is no chance of $t = \infty$ here. This also extends simply to a "capsule" shape, which is another shape that is commonly used in collision detection defined by a line segment and a radius [49], and can be thought of as a sphere that is extruded in space along that line segment.

Using simplified shapes and compound objects constructed of multiple simplified shapes allows for fast collision detection, and working in floating point allows the number system to handle the special cases. This is an instance where you can simplify the code and gain performance by allowing the number system to do what it does best.

11.7 The Gilbert–Johnson–Keerthi Algorithm

With arbitrary convex shapes, the Gilbert–Johnson–Keerthi (GJK) distance algorithm is used to find collisions by finding the minimum distance between two convex shapes [50]. GJK is computationally intensive compared to the methods mentioned before in this chapter, and works on the principle of finding the origin inside a synthetic shape that is constructed by the Minkowski difference of the two shapes being collided. Minkowski subtraction is the process of subtracting every point in a shape from every point in another shape, as shown in Figure 11.15. If the Minkowski difference contains the origin, the shapes are overlapping in space. This is too computationally expensive to use directly, but the GJK algorithm operates on this principle without computing a full Minkowski difference.

For each shape, we define a support function that finds the point that is further from the center of the shape in a given direction. For polygon geometry, when the polygon is represented as a set of vertices $P = \{\mathbf{p}_0, \ldots, \mathbf{p}_n\}$ centered at the origin, this is always one of the vertices from the set, specifically the one whose dot product with the vector pointing in that direction is maximized:

$$S(P, \mathbf{v}) = \arg\max_{\mathbf{p} \in P}(\mathbf{p} \cdot \mathbf{v})$$

For smooth objects, we instead have to find the point somewhere on the surface that maximizes this dot product. The difference between the support function of one shape in the $+\mathbf{v}$ direction and the other shape in the $-\mathbf{v}$ direction gives one of the points on the edge of the Minkowski difference of the shapes. Once we have the support function, the GJK algorithm makes use of that support function to construct geometric simplexes in Minkowski difference space to find the intersection of the shapes. A simplex is the simplest shape in a given number of dimensions that encloses space, and as such is a shape with exactly one more point than the number of dimensions: A simplex in one

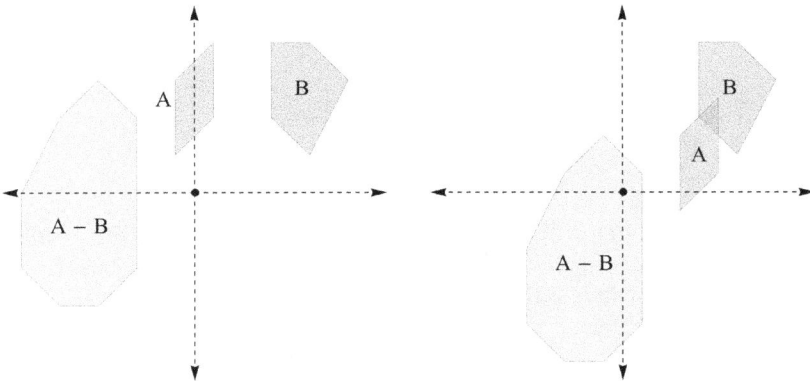

FIGURE 11.15
Two examples of Minkowski difference of shapes on cartesian plane with axes. Left: Minkowski difference of two shapes that do not collide. Right: Difference of two shapes that are overlapping.

dimension is a line segment, which extends to a triangle in two dimensions and a tetrahedron in three dimensions.

A version of the GJK algorithm for finding collisions in two dimensions is given in Algorithm 11.3. We start constructing simplexes out of the points that we get by subtracting the support functions of the two objects. At each step, if we do not cross the origin, we are done because we have gone as far as possible in that direction and there is no chance of catching the origin (see line 8). From there, we add the new support point to the simplex and check if it contains the origin. The check also tells us which direction to search next. A visual example of how the GJK algorithm works is in Figure 11.16.

The main complexity of the GJK algorithm (beyond the idea of support functions) is in handling the simplex. For the case of a simplex that does not have enough dimensions, we gather points in the simplex until we have a simplex to check (see lines 17–19). Once we have a triangle in two dimensions or a tetrahedron in three, we check for the origin behind each edge of the triangle (or equivalently, each face of the tetrahedron). If the origin is beyond one of the edges, we start a new simplex from that edge, and if we find that the origin is not behind any of the edges, we know that it is inside the simplex. An extension to three dimensions involves adding another case to the HANDLESIMPLEX function for tetrahedral simplexes that looks like the currently written case where the simplex is a triangle.

The origin cannot lie behind the vertices of the simplex during this check. When we choose support points at line 6, we then do a check (lines 7–8) to

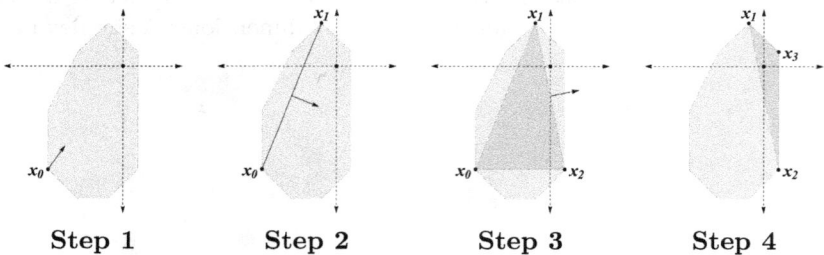

| Step 1 | Step 2 | Step 3 | Step 4 |

FIGURE 11.16

Steps of the GJK algorithm looking for a collision. Step 1: Start with an arbitrary support point, then find the direction to the origin. Step 2: Find the support point in the given direction, and using the line created between those two support points, find a normal vector pointing toward the origin. Step 3: Add the support point in that direction to create a simplex, then look for the origin, and find it past the edge from $\mathbf{x}_1$ to $\mathbf{x}_2$. Step 4: Replace $\mathbf{x}_0$ with a new support point, and find the origin in the resulting simplex, which indicates a collision.

ensure that we have crossed the origin, and if we had chosen a support point that did not cross the origin, that check would fail. At each iteration, we can also rule out the entire area behind the support points taken from the simplex in the previous iteration, because we otherwise would have chosen a different normal vector. The algorithm given here still checks the region behind that edge of the triangle.

Finally, there is a helper function given in lines 32–39 that finds a normal vector of a line segment that is facing a certain direction. This is one of several ways to do this calculation. In three (or more) dimensions, it is easy to do this using a "triple product". The vector perpendicular to a line from **a** to **b** in the direction of a point **c** is given by:

$$\mathbf{n} = [(\mathbf{b} - \mathbf{a}) \times (\mathbf{c} - \mathbf{a})] \times (\mathbf{b} - \mathbf{a})$$

The GJK algorithm extends beyond polygons by varying the support function, and can support shapes like spheres, capsules, and splines. However, these shapes that are infinitely smooth can sometimes have problems stopping and can run into numerical instability. A polygon has a discrete set of points that the support function can identify, so it is relatively easy to determine when you have searched enough simplexes. When you have an infinite number of points, you may be waiting a long time for the volume of the test simplex to converge to zero, or you might choose to stop early and sacrifice accuracy for speed. As the volume of the simplex gets smaller and smaller, numerical instability also matters: Rounding and cancellation can both cause the GJK algorithm to take steps backward. Recent efforts have improved the stability of the GJK algorithm, but it is still too computationally intensive for certain kinds of smooth surfaces to be used in situations other than those that need very high accuracy [51].

The computation time required for the GJK algorithm means that it is only used when accuracy is paramount: If drawing a number of bounding boxes around a pair of objects can also work, the GJK algorithm is a far more expensive way to check for collisions. The GJK algorithm and its family of related algorithms have one final limitation: They do not work on shapes that are in motion. The GJK algorithm detects collisions only of static objects, so you will not get a precise time during which a collision occurred during a simulation.

When speed and accuracy are both important, it is possible to combine a rough check using a set of bounding shapes with a fine collision check that uses GJK only when the rough check indicates a possible collision. This way, almost all collision checks can be rejected easily, and near-misses are still found accurately.

Algorithm 11.3 The GJK Collision Algorithm in Two Dimensions

Input: Shapes A and B and their respective support functions
Output: Boolean result indicating whether A and B have collided

1: Pick direction vector $\mathbf{d}$ ▷ Arbitrary first choice to search
2: $\mathbf{x} \leftarrow S(A, \mathbf{d}) - S(B, -\mathbf{d})$ ▷ First support point
3: Create set of points $X = \{\mathbf{x}\}$ ▷ Will hold our simplex
4: $\mathbf{d} \leftarrow -\mathbf{x}$ ▷ Point from $\mathbf{x}$ toward the origin
5: **loop**
6:      $\mathbf{x} \leftarrow S(A, \mathbf{d}) - S(B, -\mathbf{d})$ ▷ Create support points
7:      **if** $\mathbf{x} \cdot \mathbf{d} < 0$ **then**
8:          **return** FALSE   ▷ New support point did not cross origin
9:      **end if**
10:      Add $\mathbf{x}$ to X
11:      $(\mathbf{d}, r) \leftarrow$ HANDLESIMPLEX(X)
12:      **if** r **then**
13:          **return** TRUE   ▷ Origin was in the simplex
14:      **end if**
15: **end loop**

16: **procedure** HANDLESIMPLEX(X)
17:      **if** $|X| = 2$ **then**   ▷ Simplex is a line segment
18:          $\mathbf{n} \leftarrow$ NORMALFACING$(\mathbf{x}_0, \mathbf{x}_1, \mathbf{O})$ ▷ Normal toward origin
19:          **return** $(\mathbf{n}, \text{FALSE})$
20:      **else if** $|X| = 3$ **then**   ▷ Simplex is a triangle
21:          **for** i from 0 to 2 **do**
22:              $j \leftarrow (i+1) \bmod 3, \; k \leftarrow (i+2) \bmod 3$
23:              $\mathbf{n} \leftarrow -$NORMALFACING$(\mathbf{x}_i, \mathbf{x}_j, \mathbf{x}_k)$ ▷ Outward-facing normal
24:              **if** $\mathbf{n} \cdot (-\mathbf{x}_i) > 0$ **then**   ▷ Origin is in the direction of $\mathbf{n}$
25:                  Remove $\mathbf{x}_k$ from X
26:                  **return** $(\mathbf{n}, \text{FALSE})$
27:              **end if**
28:          **end for**
29:          **return** (anything, TRUE)   ▷ Origin was in the simplex
30:      **end if**
31: **end procedure**

32: **procedure** NORMALFACING$(\mathbf{a}, \mathbf{b}, \mathbf{c})$
33:      $\mathbf{t} \leftarrow \mathbf{b} - \mathbf{a}$ ▷ Vector pointing along the line segment
34:      $\mathbf{u} \leftarrow \mathbf{c} - \mathbf{a}$ ▷ Vector pointing to target
35:      $\mathbf{n} \leftarrow \left(\frac{t_y}{|\mathbf{t}|}, -\frac{t_x}{|\mathbf{t}|} \right)$ ▷ Get unit normal vector by rotating $\mathbf{t}$ 90 degrees
36:      **if** $\mathbf{n} \cdot \mathbf{u} < 0$ **then**
37:          $\mathbf{n} \leftarrow -\mathbf{n}$ ▷ Switch normals if $\mathbf{n}$ points the wrong way
38:      **end if**
39: **end procedure**

Check Your Understanding

Problem 11.1. Derive a transformation matrix for an arbitrary three-axis rotation. Perform an error analysis to determine how many ULPs of error this arbitrary rotation can induce in a vertex, assuming trigonometric functions have up to 4 ULPs of error and rotation angles are given exactly.

Problem 11.2. Create an algorithm that finds the axis-aligned bounding box of an arbitrary NURBS surface. Make a third-order NURBS surface with up to 25 control points inside the cube between $(0, 0, 0)$ and $(1, 1, 1)$ that maximizes the worst-case distance between the bounding box and the actual shape of the NURBS surface.

Problem 11.3. Bresenham's line drawing algorithm [52] is a common method for drawing straight lines using discrete pixels. Extend Bresenham's line drawing algorithm based on De Casteljau's algorithm to draw a Bézier curve. Use your algorithm to create images of curves:
A third-order curve with control points $\{(0, 3), (2, 5), (1, 1), (4, 2)\}$
A second-order curve with control points $\{(1, 2), (5, 5), (2, 4)\}$
A third-order curve with control points $\{(0, 0), (2, 2), (4, 2), (5, 1)\}$
A fourth-order curve with control points $\{(3, 0), (1, 2), (4, 3), (5, 5), (3, 0)\}$
The letter "Q" constructed out of several third-order curves

Problem 11.4. Create an algorithm for turning a third-order Bézier curve into a B-spline of the same shape, creating a knot vector and control points for the new curve. Extend your algorithm to turn a set of connected Bézier curves into a single B-spline.

Hint: This is easiest if you do not use a cardinal B-spline. Find the most convenient knot vector before working on solving for the new control points.

Problem 11.5. Derive a support function for convex NURBS splines in two-dimensional space. Build a version of the GJK algorithm for NURBS using this function. Simulate a realistic collision between two (convex) NURBS shapes. Qualitatively compare the result to a version of the collision simulator that uses a bounding box and one that uses the GJK algorithm on the polygon defined by the control points of the spline. Quantitatively compare the computing time of the three algorithms.

Challenge version: Do this for NURBS surfaces in three dimensions.

12

Numerical Integration

Numerical integration is involved in almost every video game. Any time a player needs to press a button to accelerate their character, it is backed by numerical integration. Particle simulations, fabric simulations, and fluid simulations also use numerical integration, so even completely static environments can see some use of numerical integration. It is also the backbone of most scientific simulations, which start with a given set of initial conditions and see how a series of forces interact over time on those conditions.

However, numerical integration can be a challenging piece to get right, especially with long-running physics simulations or simulations that need a high degree of accuracy. Error in numerical integration is self-compounding. Each tick depends on the values in the previous tick, so small errors accumulate over time into large problems. Quantization also introduces a new layer to integration, since we are necessarily doing addition of a small delta to a larger quantity.

12.1   Simulation Accuracy and Error

When simulating physics in floating point, we are introducing two types of discontinuity: A discrete time dimension, stepping frame by frame, and quantized space dimensions. Physics is continuous in both time and space. By discretizing in the time dimension and quantizing in the numerical dimension, we introduce two different types of errors that can both cause deviations from the real physics. Many discussions of the stability of numerical methods focus only on the physics implications of discrete time, but quantized space can also have interesting implications. In floating point, we are virtually guaranteed to have inexact calculations, and thus rounding, as part of physics simulations, because we are working with small changes in position and velocity during each timestep. A visual example of these types of error is shown in Figure 12.1.

There are two interesting ways to consider simulation accuracy, both of which can affect the aesthetics of a given simulation. These two accuracy calculations are the error of the solution at each timestep and the error in conservation of energy between timesteps. Short simulations, such as character

DOI: 10.1201/9781003565543-12

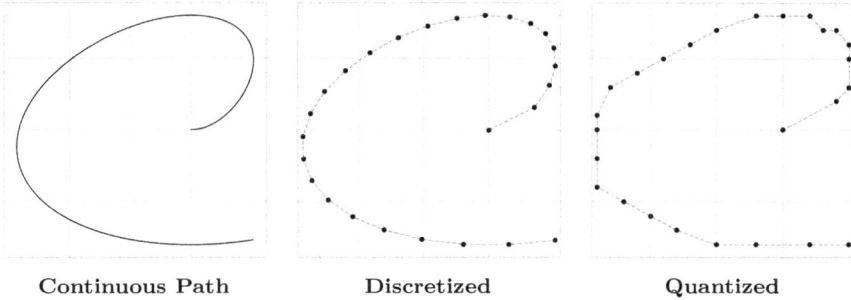

| Continuous Path | Discretized | Quantized |

FIGURE 12.1
Comparison of a motion path when discretized and quantized from the continuous version.

jumps, and simulations that "blow up" to infinity or tend toward zero may prefer to have better accuracy at the expense of conservation of energy, while long oscillating simulations will often benefit from a numerical solution that preserves conservation of energy well. However, conservation of energy may sometimes be more important, especially for long-running simulations that could end up having exponential runaway otherwise.

Discretization error is a commonly studied property of numerical systems and simulations, and is also known to numerical analysts as "truncation error", although we will refer to it as discretization error to avoid confusion since discretization error has nothing to do with rounding by truncation. Discretization error is inherent in the nature of taking discrete time steps, and naturally decreases as the size of simulation timesteps decreases. For this reason, when evaluating accuracy of integration methods, we can characterize them by the order of the discretization error they produce. The worst algorithms have discretization error that is linear in the size of the timestep, and better algorithms will have higher-order error.

A visual example of discretization error is shown in Figure 12.2 using a mass-spring system. Discretization error in simulations is path-dependent: Its future values are tightly related to its past values. This means that due to the compounding of results, discretization error tends to produce large artifacts, like significantly changing a movement path from what you might expect [53]. Discretization will also usually cause energy to fail to be conserved in a simulation, unless specifically accounted for in the algorithm used. Overshoots and undershoots caused by discretization can significantly increase or reduce the amount of total energy in a physical system by adding excess kinetic or potential energy. In the example from Figure 12.2, we can see that as the frame rate decreases (and as h increases), the simulation overshoots more and more, getting close to instability at $h = 0.6$.

However, when we introduce floating point quantization, we introduce the possibility of a small constant error term at each timestep related to

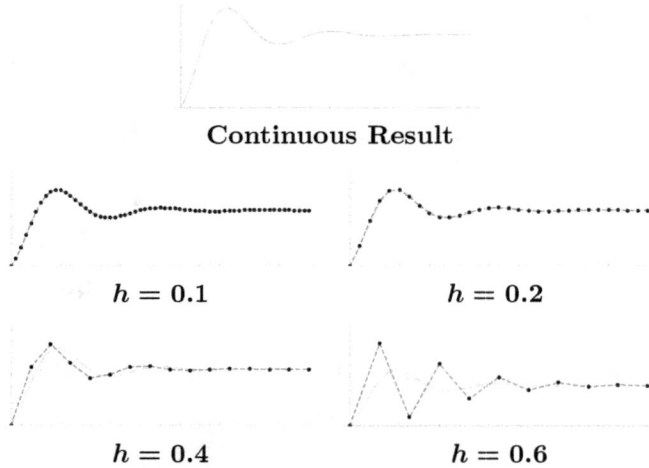

Continuous Result

$h = 0.1$ $h = 0.2$

$h = 0.4$ $h = 0.6$

FIGURE 12.2
Example of discretization error in a simulation of a mass-spring system error
with varying timestep size (h) compared to the continuous solution. The x
axis is time, and the y axis is position of the mass on the spring. Simulations
were done with leapfrog integration (see Section 12.4).

inexactness and rounding. When default rounding is used, rounding direction is essentially random when the numbers used have sufficient precision,
so quantization error will often cancel out as the simulation runs. This is not
guaranteed, though. As the precision of the calculation decreases, for example
when you put it in the distance of a scene, the calculation loses precision,
which will start to cause quantization error to become correlated with the
calculation. When this happens, the compounding blowup from quantization
error can be jarring, and can also cause instability. The normal result is that
quantization error adds some amount of Brownian motion to your simulation,
but as updates get closer to one ULP, the jitter becomes less and less random.
An example of quantization error is shown in Figure 12.3.

Interestingly, while discretization error decreases as the timestep size increases, quantization error increases. With shorter timesteps, the relative magnitude of updates decreases when compared to the values being updated. In
floating point, you can create a worse simulation by *increasing* frame rate!

A concept that is often confused with numerical accuracy is the concept of
numerical stability. Stability is a property of numerical integration algorithms
that has a set of several technical definitions, and does not say much about
accuracy in a strict sense. Many criteria for numerical stability are concerned
with what happens when time goes to infinity. In other words, if your simulation always ends up reaching 0 or infinity correctly, it's stable, no matter the
path it took to get there. Quantization can also cause instability by moving
past points where the system would otherwise be stationary.

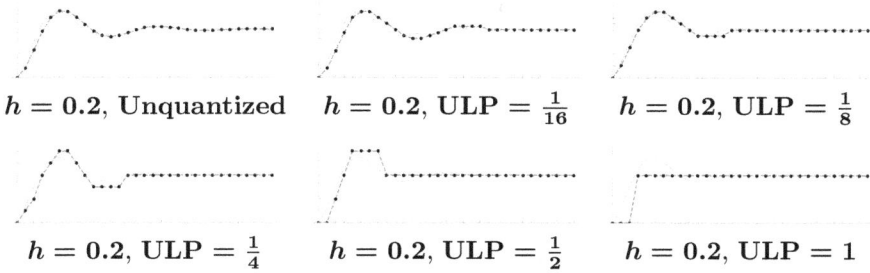

$h = 0.2$, Unquantized $\qquad$ $h = 0.2$, ULP $= \frac{1}{16}$ $\qquad$ $h = 0.2$, ULP $= \frac{1}{8}$

$h = 0.2$, ULP $= \frac{1}{4}$ $\qquad$ $h = 0.2$, ULP $= \frac{1}{2}$ $\qquad$ $h = 0.2$, ULP $= 1$

FIGURE 12.3
Examples of quantization error in a simulation of a mass-spring system error with $h = 0.2$ compared to the unquantized version, starting with five significant bits (ULP of $\frac{1}{16}$ times the end state size) and going to 1.

12.2 Euler Integration

Euler integration (Euler's method) is the naive approach to numerical integration, so we will discuss a general version before thinking about a physics-specific version [54]. Given a first-order vector-valued differential equation:

$$\mathbf{y}' = \mathbf{f}(t, \mathbf{y})$$

Euler's method computes a discretized version of $\mathbf{y}$ by evaluating $\mathbf{f}$ at each chosen point, moving straight down that path for a small amount of time, h, and then repeating:

$$\mathbf{y}_{i+1} = \mathbf{y}_i + \mathbf{f}(t, \mathbf{y}_i)h \qquad (12.1)$$

Essentially, Euler's method takes the derivative in the differential equation and makes it represent the slope of the system for a small, finite step rather than having it represent an instantaneous slope. As h goes to 0, the iteration of Euler's method converges to the original differential equation. Euler's method has the power to also be used on higher-order scalar differential equations by using a vector of derivatives. Taking a function of the form:

$$a^{(n+1)}(t) = f\left(t, a, a', a'', ..., a^{(n)}\right)$$

We write the higher-order differential equation in the following form:

$$\begin{bmatrix} a' \\ a'' \\ \vdots \\ a^{(n+1)} \end{bmatrix} = \begin{bmatrix} a' \\ a'' \\ \vdots \\ f\left(t, a, a', a'', ..., a^{(n)}\right) \end{bmatrix}$$

And then discretize that system of equations using the method we saw previously:

$$
\begin{bmatrix} a_{i+1} \\ a'_{i+1} \\ \vdots \\ a^{(n)}_{i+1} \end{bmatrix} = \begin{bmatrix} a_i + a'_i h \\ a'_i + a''_i h \\ \vdots \\ a^{(n)}_i + f\left(t, a, a', a'', ..., a^{(n)}\right) h \end{bmatrix}
$$

In physical systems, the first equation covers the change in position over time, the second equation the change of velocity, and so on. For games, the system of equations often stops at position and velocity. This means that Euler's method as used in games is usually written as:

$$
x_{i+1} = x_i + f(x_i, v_i, t)h
$$
$$
v_{i+1} = v_i + g(x_i, v_i, t)h
$$

where $g(x_i, v_i, t)$ is the entity's acceleration, which is a function of their position and a combination of the forces acting on the entity and the user's inputs (especially for players) over time, and the function $f(x_i, v_i, t)$ is usually just the entity's velocity, but can have some modifications for collision detection or other effects. Most objects in many games are affected by gravity, the player has some acceleration from the controller, and other forces like object collisions can also accelerate an entity. In reality, most games will have two or three position and velocity dimensions, meaning a system of 4 or 6 equations, but the relation between the equations holds the same as for one dimension.

Continuing with the example of Figure 12.2, damped mass-spring system, we have the following f and g:

$$
f(x, v, t) = v
$$
$$
g(x, v, t) = \frac{F_s}{m} - \frac{F_f}{m} = \frac{k}{m}(1 - x) - \frac{v}{m}
$$

The f function is straightforward. The first term of the g function handles the spring force, with the spring constant $k = 5$, and a mass, m of 0.5. The second term is a damping function that models friction in the mass-spring system, whose value is dependent on v, with a damping constant of $\frac{1}{m}$. This system, in physical terms, is an underdamped mass-spring system. It settles at $x = 1$, but overshoots a bit and oscillates with a period of $\frac{2\pi}{3} \approx 2$. You may notice that g is dependent on velocity. This form works for Euler's method, but we will need to do some algebra for later methods.

The results of Euler integration of this system are shown in Figure 12.4. With $h = 0.01$, equivalent to 100 FPS if we take one unit of time as one second, this simulation is okay, and lines up with the real system. With $h = 0.05$, we start to see some divergence, as the Euler system clearly has extra energy. When we get to $h = 0.1$, there is significant divergence, and with $h = 0.2$, we

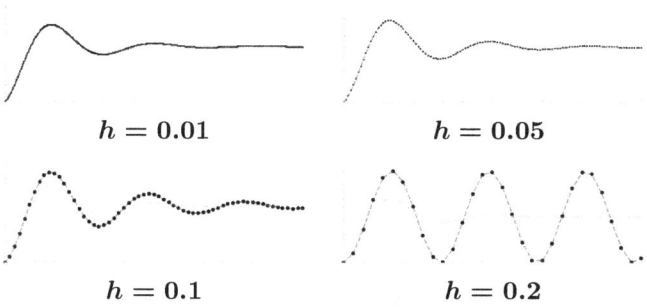

FIGURE 12.4
Euler integration of the mass-spring system from Figure 12.2, compared against the analytical solution. Even at $h = 0.1$, we get significant deviation, and we are unstable by $h = 0.2$.

are already unstable. However, quantization hits hard when you have to use such small timesteps. You will necessarily be adding small quantities to large quantities, so your updates are at risk of cancellation. Figure 12.5 demonstrates Euler's method with quantization, showing the importance of having enough precision for smaller timesteps. As is shown in the figure, rounding causes energy conservation issues on its own, with the simulations at 6 bits of significance having faster motion and wrong final settling points.

While Euler's method is a classic and it is both easy and fast, there are few circumstances in which you should be using Euler's method. Other methods with better numerical properties are equally fast and equally simple. Keeping Euler's method stable is often too much of a chore.

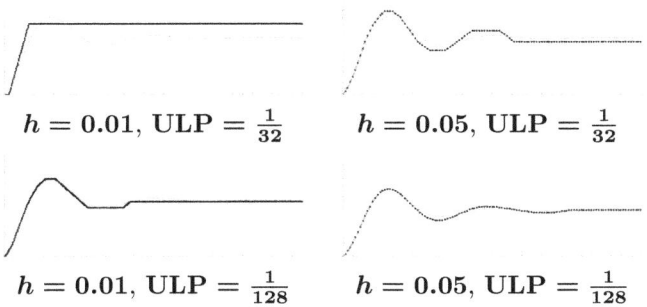

FIGURE 12.5
Euler integration of the mass-spring system from Figure 12.2 with rounding. The simulations with $h = 0.01$ are less accurate than the ones with $h = 0.05$. Both simulations with 6 bits of significance settle at the wrong final position.

12.3 Adding Energy Conservation to Euler

One of the many foibles of Euler's method is that it does not conserve energy. However, with a slight modification, Euler's method becomes energy-conserving in some limited circumstances. The modified algorithm is known as **semi-implicit Euler**, and only applies to second-order physical systems (more rigorously, to Hamiltonian systems). Unlike Euler's method, it does not generalize to systems of general differential equations. Thankfully, game engines are built on second-order physics simulations. The modification is simple and clever: Instead of updating position and velocity simultaneously, we compute one of them forward ("explicitly") and we compute the other one looking backward ("implicitly") [55]. In equation form the two alternatives are:

$$x_{i+1} = x_i + f(v_i, t)h \qquad v_{i+1} = v_i + g(x_i, t)h$$
$$v_{i+1} = v_i + g(x_{i+1}, t)h \qquad x_{i+1} = x_i + f(v_{i+1}, t)h \tag{12.2}$$

The left alternative does an implicit computation of velocity, while the left side does an implicit computation of position. Implicit velocity is the more common case, and usually the more useful one.

This may seem like a simple change, but we are now staggering the computation of velocity and position, either using the old velocity to compute the new position and then computing the new velocity with the new position, or using the old position to compute velocity and using that velocity to compute a new position. The forward-looking update adds energy to the system, and the backwards-looking update subtracts a corresponding amount of energy. The position update operates only on potential energy and the velocity update operates only on kinetic energy. Semi-implicit Euler with implicit velocity will thus over-estimate potential energy and then under-estimate the resulting kinetic energy.

The mathematical term for energy-conserving approximations is that they are **symplectic**. Symplectic methods do not actually fully conserve energy, but they do conserve a relation that approximates energy. The rigorous derivation of the concept of a symplectic integrator depends on Hamiltonian mechanics, but we do not need it to get an intuitive understanding. Energy conservation is approximately correct to the order of the approximation. Since semi-implicit Euler is a first-order algorithm, it will conserve energy to first order, only successfully conserving energy when energy is constant.

Applying semi-implicit Euler to our mass-spring system may need us to do a bit of algebra. Since our g function is a function of velocity, we need to eliminate velocity by plugging v_{i+1} into g and canceling:

$$v_{i+1} = v_i + g(x_{i+1}, v_{i+1}, t)h = v_i + \frac{h}{m}\left[k(1 - x_{i+1}) - v_{i+1}\right]$$

Combining the v_{i+1} terms,

$$v_{i+1}\left(1 + \frac{h}{m}\right) = v_i + \frac{hk(1 - x_{i+1})}{m}$$

Moving everything to the right side, we get:

$$v_{i+1} = \frac{v_i + \frac{hk(1-x_{i+1})}{m}}{1 + \frac{h}{m}} = \frac{v_i}{1 + \frac{h}{m}} + \frac{k(1 - x_{i+1})}{1 + \frac{m}{h}}$$

Semi-implicit Euler preserves conservation of energy in circumstances when the forces applied to the object do not change over time. It is not a big deal that there is no simple "$v_i + ...$" term on the right side of the equation, all that matters is that v_{i+1} varies based on position. However, the other formulation of semi-implicit Euler, computing velocity first, is okay without rearrangement. The computational complexity of this iteration is not bad since most of the terms being multiplied and divided here are constants:

$$v_{i+1} = \xi_1 v_i + \xi_2(1 - x_{i+1})$$

This can be computed with a subtraction, a multiplication, and an FMA. The full iteration has an additional FMA to update position. This simple change in Euler's method gives us much better simulation results, demonstrated in Figure 12.6.

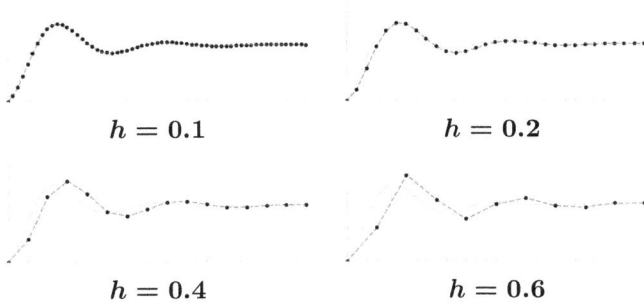

$h = 0.1$ $h = 0.2$

$h = 0.4$ $h = 0.6$

FIGURE 12.6
Semi-implicit Euler integration of the mass-spring system from Figure 12.2 with implicit velocity, compared against the analytical solution. We retain stability even with $h = 0.6$, and stay relatively accurate even with a large simulation timestep.

Thanks to the conservation of energy, we retain stability longer than any previous method shown, despite overshooting. The one drawback is the need to keep simulation equations in this type of canonical form. However, we still needed to do some algebra here when we have friction forces and other forces based on the velocity of the object being simulated, making simulator engines substantially harder to write when these forces are present. To overcome this, many graphics simulations use a form of dynamics that is based entirely on position, known as "position-based dynamics" [56]. In this case, the algebraic manipulation we had to do is built into the physics system.

It is also possible to calculate position implicitly while performing a forward calculation of velocity. For this mass-spring system, it is much simpler to do so because we don't need to run the velocity calculation. This simulation would also be a semi-implicit Euler calculation:

$$v_{i+1} = v_i + \frac{h}{m} \left[k(1 - x_i) - v_i \right]$$
$$x_{i+1} = x_i + v_{i+1}h$$

This is a far less common form of calculation, since the algebra we did on velocity now must be done for position, which can be more complicated when considering simulations that have collisions or other forces that are position-dependent. For completeness, simulations with this method are shown in Figure 12.7. Especially with position-based dynamics, this form of semi-implicit Euler is not usually used in practice.

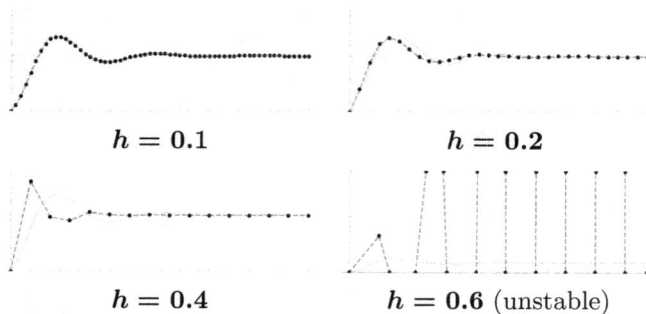

$h = 0.1$ $h = 0.2$

$h = 0.4$ $h = 0.6$ (unstable)

FIGURE 12.7
Semi-implicit Euler integration of the mass-spring system from Figure 12.2 with implicit position, compared against the analytical solution. While the previous simulation "dragged" behind the real system, this version runs ahead, and becomes unstable on this system at a lower h.

12.4   Leapfrog Integration and Velocity Verlet

Leapfrog integration improves on Euler's method with a simple idea that takes the idea of semi-implicit Euler one step further. We update velocity on synthetic half-frames and update position on the whole-number frames. The iteration loop is:

$$v_{i+\frac{1}{2}} = v_{i-\frac{1}{2}} + g(x_i, t)h$$
$$x_{i+1} = x_i + v_{i+\frac{1}{2}}h \tag{12.3}$$

If g depends on v, we need to compute it implicitly at $v = v_i$. In order to substitute v_i for g in our mass-spring system, we use a linear interpolation of the velocity halfway between the step we have already calculated and the step we are about to calculate:

$$v_i = \frac{v_{i+\frac{1}{2}} + v_{i-\frac{1}{2}}}{2}$$

Then we need to do exactly what we did in semi-implicit Euler to isolate the v_i terms:

$$v_{i+\frac{1}{2}} = v_{i-\frac{1}{2}} + \frac{hk}{m}(1 - x_i) - \frac{h}{m}\left[\frac{v_{i+\frac{1}{2}} + v_{i-\frac{1}{2}}}{2}\right] = \frac{1 - \frac{h}{2m}}{1 + \frac{h}{2m}} v_{i-\frac{1}{2}} + \frac{hk(1 - x_i)}{m + \frac{h}{2}}$$

Canonically, the entire acceleration update rule goes into the velocity calculation, while the position update rule becomes a simple move by the calculated velocity. However, we need to set an initial value for $v_{-\frac{1}{2}}$ or compute $v_{\frac{1}{2}}$, which is not necessarily equal to v_0. The typical method for doing this is to do a half-sized round of Euler simply to find $v_{\frac{1}{2}}$, and then move on to leapfrog integration for the rest of the simulation. Again, despite looking complicated, most of the complexity is in the form of constants, and the compute time is exactly the same as semi-implicit Euler.

Despite the computation time being the same as semi-implicit Euler, leapfrog integration is a second-order method. Its accuracy is quite a bit better than semi-implicit Euler in most cases. Leapfrog integration is also symplectic, and behaves well for simulations that rely heavily on conservation of energy.

Figure 12.2 from the beginning of the chapter was generated using leapfrog integration. It is our first second-order method, and as such has better discretization error scaling than the first-order methods we have discussed. Comparing Figure 12.2 to Figure 12.6, it looks like we are actually doing worse than semi-implicit Euler, given the inaccuracy with $h = 0.6$, but we are looking at the long-range behavior of the simulator in a periodic system. What we see here is actually that semi-implicit Euler with implicit velocity actually has better behavior on periodic systems with large h, due to the fact that velocity of semi-implicit Euler runs a step ahead, while leapfrog integration only runs half a step ahead. Leapfrog integration of periodic systems is stable only when $h < \frac{2}{\omega}$ where ω is the angular frequency of the oscillation (in this system, $\omega = 3$), and since we are simulating with double precision, we expect stability to break at a slightly lower h.

If we look at the error of the simulation at a high step rate, we get a comparison like Figure 12.8. This shows a significant improvement of accuracy compared to semi-implicit Euler, as would be expected from a second-order method when compared to a first-order method. When we consider quantization, the errors look like Figure 12.9. With enough bits of significance, the higher-order method performs as expected, but note that the shape of the error is no longer smooth. As the significance level decreases, the difference in accuracy between semi-implicit Euler and the leapfrog integration method decreases, to the point where both reach a wrong settling point and have comparable levels of error on small motions. The higher-order integrator still performs better when the simulation moves quickly, but as the dynamics level out, the error from quantization becomes dominant over the error from discretization.

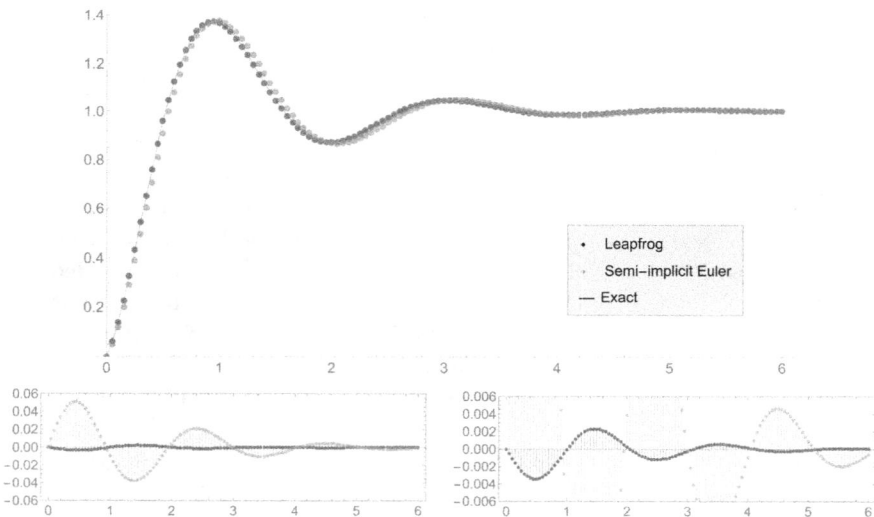

FIGURE 12.8
Simulation results using leapfrog integration and semi-implicit Euler using $h = 0.05$, with two plots of approximation error at different scales. The black line is exact, dark gray dots are leapfrog, and light gray dots are semi-implicit Euler.

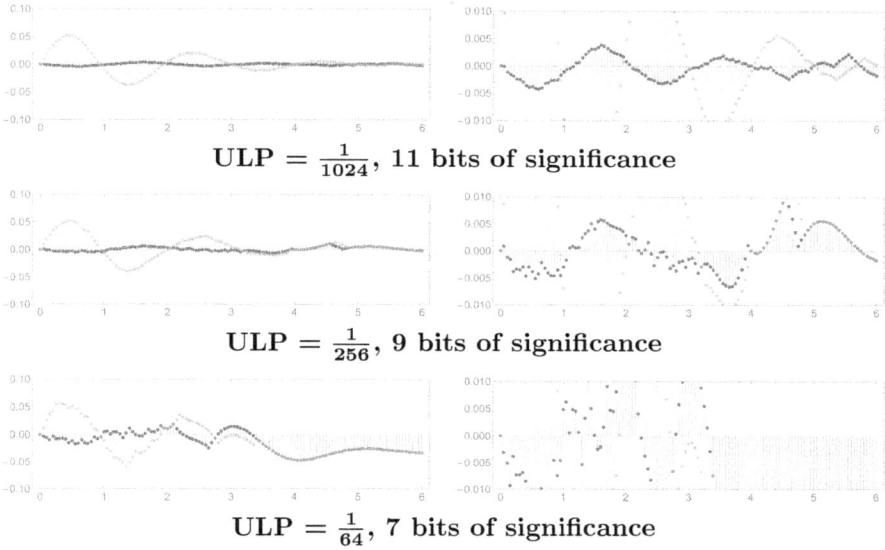

$$\text{ULP} = \tfrac{1}{1024}, \text{ 11 bits of significance}$$

$$\text{ULP} = \tfrac{1}{256}, \text{ 9 bits of significance}$$

$$\text{ULP} = \tfrac{1}{64}, \text{ 7 bits of significance}$$

FIGURE 12.9
Comparison of simulation error for semi-implicit Euler and leapfrog integration with $h = 0.05$ and varying quantization levels, showing the jitter effect from rounding. The simulation with only 7 bits of significance has similar performance for the two algorithms, and also has a wrong settling point in both algorithms.

Instead of doing this level of algebra, it is possible to split the velocity calculation during leapfrog integration into an iteration with three steps:

$$v_{i+\frac{1}{2}} = v_i + g(x_i, v_i, t)\tfrac{h}{2}$$
$$x_{i+1} = x_i + v_{i+\frac{1}{2}} h \qquad (12.4)$$
$$v_{i+1} = v_{i+\frac{1}{2}} + g(x_{i+1}, v_{i+1}, t + h)\tfrac{h}{2}$$

While more computationally intensive at runtime, this allows us to use g as a function of v_i in a straightforward way, and it allows the method to work with a non-constant timestep. It has the somewhat ferocious name of the **kick-drift-kick** formulation of leapfrog integration. In this case, acceleration feeds to velocity explicitly in the first velocity update step, and implicitly in the second update step. Practically, this allows the second update step to look like semi-implicit Euler when acceleration depends on velocity. This formulation is also a form of **velocity Verlet** integration, which is one of the most common forms of numerical integration in game engines today.

Unlike leapfrog integration, which combines these three steps by keeping the velocity step staggered with the position step, the common form of velocity Verlet accepts a second-order term in the position calculation to keep the updates of position and velocity aligned in time:

$$x_{i+1} = x_i + v_i h + \tfrac{1}{2} g(x_i, t) h^2$$
$$v_{i+1} = v_i + \frac{g(x_i, t) + g(x_{i+1}, t+h)}{2} h \qquad (12.5)$$

The two-equation form of Verlet assumes that g is independent of velocity, and there is no easy way in this expression form to perform the transformations we have before to remove velocity from the equation. This form also assumes a timestep of a constant size, while the kick-drift-kick form does not require a constant timestep. Some simulation engines will use the kick-drift-kick form for this reason, despite the fact that it is slightly more computationally expensive. However, the two-equation Verlet formulation is a favorite in game engines that can use a constant simulation frame rate and prefer the higher-performance variant.

Hacking Velocity-dependent Dynamics into Velocity Verlet

Implementations of velocity Verlet will sometimes add velocity-dependence to g by using v_i and adding a straightforward "acceleration" step between the updates of position and velocity instead of depending on v_{i+1} [57].

This addition will significantly compromise the numerical accuracy of the integrator and break its equivalence with leapfrog integration, as well as breaking energy conservation. For the kick-drift-kick form:

$$v_{i+\frac{1}{2}} = v_i + a_i, t\tfrac{h}{2}$$
$$x_{i+1} = x_i + v_{i+\frac{1}{2}} h$$
$$a_{i+1} = g(x_{i+1}, v_{i+\frac{1}{2}}, t+h)$$
$$v_{i+1} = v_{i+\frac{1}{2}} + a_{i+1}\tfrac{h}{2}$$

where we accept a velocity half a step behind for the acceleration calculation, and for the simplified velocity Verlet form we accept a velocity a full step behind:

$$x_{i+1} = x_i + v_i h + \tfrac{1}{2} a_i h^2$$
$$a_{i+1} = g(x_{i+1}, v_i, t+h)$$
$$v_{i+1} = v_i + \frac{a_i + a_{i+1}}{2} h \qquad (12.6)$$

Both of these are compromises of numerical accuracy against the avoidance of a need to do algebra, and the compromise is significant. In each of these cases, the acceleration is calculated as a straightforward computation based on the forces on the object just like in Euler's method. However, using properly implemented leapfrog integration or semi-implicit Euler will usually give better results, and this actually transforms the symplectic second-order method of velocity Verlet into a non-energy-conserving first-order method.

12.5   Predictor-Corrector Methods

Up to now, we have been refining Euler's method by altering the times at which we sample and calculate things. Semi-implicit Euler performs a backwards calculation and a forwards calculation, and velocity Verlet takes velocity at the midpoint between position updates. Another high-level approach we can take is that we can spend longer calculating to use a more accurate estimate of the slope by making and refining our predictions. The general class of algorithms to do this has the name **predictor-corrector methods**, since these methods make several predictions and self-correct each time.

We saw with the leapfrog method that taking velocity samples at the midpoints between position samples gives us significant improvements in accuracy. Going back to the original Euler formulation:

$$\mathbf{y}' = \mathbf{f}(t, \mathbf{y}) \quad \longrightarrow \quad \mathbf{y}_{i+1} = \mathbf{y}_i + \mathbf{f}(t, \mathbf{y}_i)h$$

We would prefer to compute:

$$\mathbf{y}_{i+1} = \mathbf{y}_i + \mathbf{f}\left(t + \tfrac{h}{2}, \mathbf{y}_{i+\frac{1}{2}}\right)h$$

As with leapfrog integration, this approach should square our error and produce a second-order method out of our original first-order Euler method. The problem is that we don't know $\mathbf{y}_{i+\frac{1}{2}}$ exactly. However, we can just use Euler's method to estimate it:

$$\mathbf{y}_{i+\frac{1}{2}} \approx \mathbf{y}_i + \mathbf{f}\left(t, \mathbf{y}_i\right)\tfrac{h}{2}$$

That gives us the **midpoint method** for numerical integration:

$$\mathbf{y}_{i+1} = \mathbf{y}_i + \mathbf{f}\left(t + \tfrac{h}{2}, \mathbf{y}_i + \mathbf{f}\left(t, \mathbf{y}_i\right)\tfrac{h}{2}\right)h \tag{12.7}$$

At each step, we first take half an Euler step forward, but then instead of using that result, we evaluate $\mathbf{f}$, the slope of the path of $\mathbf{y}$, at that point. We when discard that Euler step and add our newly computed slope to $\mathbf{y}_i$. This allows us to take our slope calculation at a pseudo-midpoint between $\mathbf{y}_i$ and $\mathbf{y}_{i+1}$.

We have another option to get our midpoint slope, which is to average the slopes at $\mathbf{y}_i$ and $\mathbf{y}_{i+1}$:

$$\mathbf{f}\left(t + \tfrac{h}{2}, \mathbf{y}_{i+\frac{1}{2}}\right) \approx \frac{\mathbf{f}\left(t, \mathbf{y}_i\right) + \mathbf{f}\left(t + h, \mathbf{y}_{i+1}\right)}{2}$$

We still have the same problem, which is that we don't know a_i. We can just estimate it, however, using Euler's method. We will then discard the result, as with the midpoint method, and use the estimated future value of $\mathbf{y}$ to get a better iteration step:

$$\tilde{\mathbf{y}}_{i+1} = \mathbf{y}_i + \mathbf{f}(t, \mathbf{y}_i)h$$

$$\mathbf{y}_{i+1} = \mathbf{y}_i + \frac{h}{2}\left[\mathbf{f}(t, \mathbf{y}_i) + \mathbf{f}(t + h, \tilde{\mathbf{y}}_{i+1})\right] \tag{12.8}$$

This method is called **Heun's method**, after the mathematician who invented it, and is also called the "trapezoid rule" in connection to a method of integrating functions of known value. Heun's method can be alternatively formulated as a combination of Euler's method (looking forward) and then the trapezoid rule (looking backward).

Both of these methods build on Euler's method by using Euler's method to find a midpoint value, either of **y** or **f**, and plugging that in as our new increment for the step. Incidentally, both methods are known as the "modified Euler method" to different people. A graphical representation of what is happening in each of these methods is shown in Figure 12.10.

Interestingly, on the mass-spring system we have been working with, Heun's method and the midpoint method produce exactly the same calculation. This will be true for several physics systems, especially with time-independent dynamics. In this case, midpoint integration uses the following formula for position:

$$x_{i+1} = x_i + v_{i+\frac{1}{2}}h = x_i + \left[v_i + \tfrac{h}{2}g(x_i, v_i)\right]h$$

We can rearrange this a bit to see that it is in fact equal to the Heun's method formula:

$$x_{i+1} = x_i + \tfrac{h}{2}\left[2v_i + g(x_i, v_i)h\right] = x_i + \tfrac{h}{2}\left[v_i + v_i + g(x_i, v_i)h\right] = x_i + \tfrac{h}{2}\left[v_i + \tilde{v}_{i+1}\right]$$

We can see that we just have $v_i + \tilde{v}_{i+1}$ in the parentheses. A big difference between midpoint and Heun comes from time dependence, but there is no time dependence here. A similar transformation happens with the velocity formula. Several example simulations of this system with midpoint/Heun integration are shown in Figure 12.11. The result is nothing special, but falls in the same class as velocity Verlet. It is reasonably accurate, but struggles with the periodic behavior as h increases. You can also see in the figure that for higher

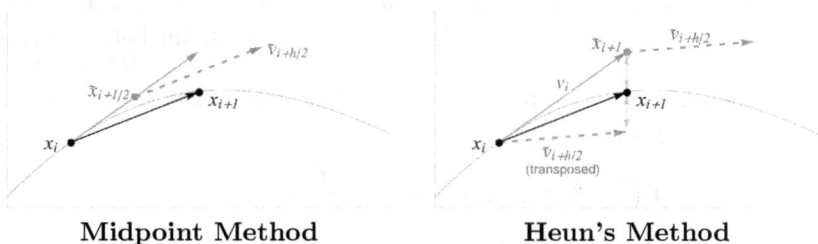

Midpoint Method   **Heun's Method**

FIGURE 12.10
Geometric approaches to second-order predictor-corrector methods. The midpoint method advances up to the midpoint of the Euler step (solid gray arrow) to find the derivative at the midpoint, which is used to update position. Heun's method advances to the end of the Euler step to find the "next" derivative, and averages the initial and final derivatives.

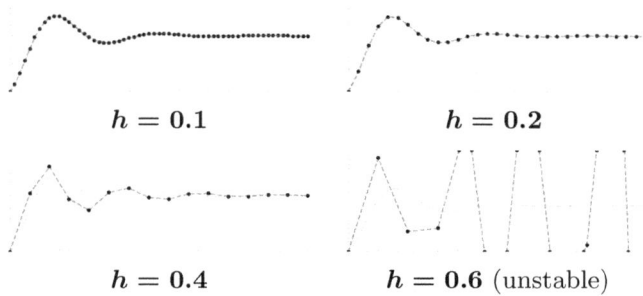

$h = 0.1$ $h = 0.2$

$h = 0.4$ $h = 0.6$ (unstable)

FIGURE 12.11
Midpoint/Heun Euler integration of the mass-spring system from Figure 12.2, compared against the analytical solution. Note that the frequency of oscillation increasing as h increases.

h, the frequency of oscillation increases: At $h = 0.2$, the midpoint method oscillates slightly faster than the system, and as of $h = 0.4$, the frequency difference is prominent. This is due to the fact that midpoint integration adds energy to the system, just like Euler integration. We have left the realm of symplectic integrators.

Midpoint integration is also somewhat slower than Euler. We effectively do two Euler rounds for each simulation step. This will also make midpoint integration slower than leapfrog integration and even possibly slower than the velocity Verlet formulations. The benefit of midpoint integration is that it is not specialized to second-order systems like those alternatives.

12.6   Runge–Kutta Methods

We have done one round of prediction and correction, so why stop there? The most popular method for simulating differential equations doesn't: It goes through another two rounds of correction. Our original motivation was to find the derivative in the middle of $\mathbf{y}_i$ and $\mathbf{y}_{i+1}$, instead of using the derivative at the start, to find a good slope between the two points. However, it doesn't address the curvature of the space between the two points well. That curvature is the realm of the higher-order derivatives.

To motivate the Runge–Kutta methods, think about the Taylor series of $\mathbf{y}(t)$:

$$\mathbf{y}(t + h) = \mathbf{y}(t) + h\mathbf{y}'(t) + \tfrac{h^2}{2}\mathbf{y}''(t) + \tfrac{h^3}{6}\mathbf{y}'''(t) + \cdots$$

We know $\mathbf{y}(t)$, and our midpoint-based estimators are good at getting a first derivative that is as close to $h\mathbf{y}'(t) + \tfrac{h^2}{2}\mathbf{y}''(t) + \cdots$ as possible. However, if we keep testing the error of our approximation, we can start to account for higher-order variance. To get there, we will have to do some calculus when we substitute based on $\mathbf{y}' = \mathbf{f}(t, \mathbf{y})$. If we expand our Taylor series, the first two terms are simple, but later terms need to use the chain rule since f depends on both t and $\mathbf{y}(t)$, and they get very complicated:

$$\mathbf{y}(t + h) = \mathbf{y}(t) + h\mathbf{f} + \frac{h^2}{2}\left[\frac{d\mathbf{f}}{dt} + \frac{d\mathbf{f}}{d\mathbf{y}}\mathbf{f}\right]$$

$$+ \frac{h^2}{6}\left[\frac{d^2\mathbf{f}}{dt^2} + \left(\frac{d\mathbf{f}}{d\mathbf{y}}\right)^2\mathbf{f} + \frac{d^2\mathbf{f}}{d\mathbf{y}^2}\mathbf{f}^2 + \frac{d\mathbf{f}}{dt}\frac{d\mathbf{f}}{d\mathbf{y}} + 2\left(\frac{d}{dt}\frac{d\mathbf{f}}{d\mathbf{y}}\right)\mathbf{f}\right]$$

$$+ \cdots$$

If we want to search further derivatives of the path of $\mathbf{y}$, we need to be able to probe the field of $\mathbf{f}$ to get a combination of derivatives that line up with parts of this Taylor series.

As before, we will be probing by making predictions $(k_1, k_2, ..., k_n)$ and stepping forward on our predictions to evaluate the next direction to probe. In a general form, the first two predictions we can make are a generalization of the midpoint method:

$$k_1 = \mathbf{f}(t_i, \mathbf{y}_i) \qquad \text{(Euler)}$$
$$k_2 = \mathbf{f}(t_i + c_2 h, \mathbf{y}_i + a_{21}k_1 h) \qquad \text{(Generalized Midpoint)}$$

We first try the Euler vector. Every algorithm we have done has started here, and there are no parametric constants to deal with. The next thing we do is take k_2 as the vector at some point down the Euler vector, parametrized by c_2 and a_{21}. a_{21} dictates how far down we go on k_1 in $\mathbf{y}$ space, and c_2 dictates how much time we advance. Usually, we will have $c_2 = a_{21}$, but in a generic form, we leave the constraint free. We can go further by defining a third probe:

$$k_3 = \mathbf{f}(t_i + c_3 h, \mathbf{y}_i + [a_{31}k_1 + a_{32}k_2]h)$$

Keeping things generic, our third probe line goes forward in time by c_3 and advances down some linear combination of k_1 and k_2, parametrized by a_{31} and a_{32} respectively. Now that we have the idea, there's no reason we can't go further:

$$k_4 = \mathbf{f}(t_i + c_4 h, \mathbf{y}_i + [a_{41}k_1 + a_{42}k_2 + a_{43}k_3]h)$$
$$k_5 = \mathbf{f}(t_i + c_5 h, \mathbf{y}_i + [a_{51}k_1 + a_{52}k_2 + a_{53}k_3 + a_{54}k_4]h)$$

$$\vdots$$

$$k_n = \mathbf{f}\left(t_i + c_n, \mathbf{y}_i + h\sum_{m=1}^{n-1} a_{nm}k_m\right)$$

At each step, we use the information from the previous probes to determine where to aim next, and sample that vector. Once we have all of our predictions, we can produce a new value of $\mathbf{y}$ as a weighted average of those predictions:

$$\mathbf{y}_{i+1} = \mathbf{y}_i + h \sum_{i=1}^{n} b_i k_i$$

Now the only thing that remains is to solve for all the a's, b's and c's to produce a useful method. If you got through all that, you have finally reached the hard part! This is generally done by solving for a's, b's, and c's with a large system of equations to match some number of terms of the Taylor series from above. Methods of this generic form are known as **Runge–Kutta** methods (RK for short), after the two mathematicians who described the framework and first applied it.

Just for fun, let's solve for a second-order Runge–Kutta method. To keep our error third-order or worse we can put the Runge–Kutta solution on the left and the second-order Taylor expansion of $\mathbf{y}$ on the right side of the equals sign:

$$\mathbf{y}_i + h(b_1 k_1 + b_2 k_2) = \mathbf{y}_i + h\mathbf{f}(t_i, \mathbf{y}_i) + \frac{h^2}{2} \left[\frac{d\mathbf{f}}{dt} + \frac{d\mathbf{f}}{d\mathbf{y}} \mathbf{f} \right]$$

Simplifying a bit by dropping the constant terms (and putting the derivatives into nicer notation):

$$h(b_1 k_1 + b_2 k_2) = h\mathbf{f}(t_i, \mathbf{y}_i) + \tfrac{h^2}{2} \left[\mathbf{f}_t(t_i, \mathbf{y}_i) + \mathbf{f}_\mathbf{y}(t_i, \mathbf{y}_i)\mathbf{f}(t_i, \mathbf{y}_i) \right]$$

The k_2 term will not expand nicely if we try to just substitute it—k_2 depends on $\mathbf{f}$ evaluated at a point dependent on $\mathbf{f}$, and doing that evaluation will not give us a generic method. However, we have already started taking Taylor series and dropping high-order terms, so let's take a first-order Taylor series of k_2 around $h = 0$:

$$k_2 = \mathbf{f}(t_i, \mathbf{y}_i) + h \left[c_2 \mathbf{f}_t(t_i, \mathbf{y}_i) + a_{21} \mathbf{f}_\mathbf{y}(t_i, \mathbf{y}_i)\mathbf{f}(t_i, \mathbf{y}_i) \right]$$

Since we're going to multiply this series by h, we only need a first-order series to get a second-order approximation on the left side of the equals sign. $k_1 = h\mathbf{f}(t_i, \mathbf{y}_i)$ is also easy to put in exactly without approximating. This gives us a new equation to solve:

$$h\mathbf{f}(t_i, \mathbf{y}_i) [b_1 + b_2] + h^2 [b_2 c_2 \mathbf{f}_t(t_i, \mathbf{y}_i) + b_2 a_{21} \mathbf{f}_\mathbf{y}(t_i, \mathbf{y}_i)\mathbf{f}(t_i, \mathbf{y}_i)]$$
$$= h\mathbf{f}(t_i, \mathbf{y}_i) + \tfrac{h^2}{2} [\mathbf{f}_t(t_i, \mathbf{y}_i) + \mathbf{f}_\mathbf{y}(t_i, \mathbf{y}_i)\mathbf{f}(t_i, \mathbf{y}_i)]$$

Isolating the h part and the h^2 part, we get a system of two equations with four variables:

$$\mathbf{f}(t_i, \mathbf{y}_i) [b_1 + b_2] = \mathbf{f}(t_i, \mathbf{y}_i)$$
$$b_2 c_2 \mathbf{f}_t(t_i, \mathbf{y}_i) + b_2 a_{21} \mathbf{f}_\mathbf{y}(t_i, \mathbf{y}_i)\mathbf{f}(t_i, \mathbf{y}_i) = \tfrac{1}{2} [\mathbf{f}_t(t_i, \mathbf{y}_i) + \mathbf{f}_\mathbf{y}(t_i, \mathbf{y}_i)\mathbf{f}(t_i, \mathbf{y}_i)]$$

That gives us four simple rules to make a second-order Runge–Kutta method:

$$b_1 + b_2 = 1 \qquad \text{(from linear term; coefficient of } \mathbf{f})$$
$$b_2 c_2 = \tfrac{1}{2} \qquad \text{(from quadratic term; coefficient of } \mathbf{f}_t)$$
$$b_2 a_{21} = \tfrac{1}{2} \qquad \text{(from quadratic term; coefficient of } \mathbf{f}_\mathbf{y}\mathbf{f})$$
$$b_2 \neq 0 \qquad \text{(force nonzero second-order term)}$$

We can choose any set of coefficients we would like within these rules, although it helps if all of the c and a terms are positive so that we are actually making predictions.

Let's choose $b_2 = 1$, $b_1 = 0$, $c_2 = a_{21} = \tfrac{1}{2}$:

$$k_2 = \mathbf{f}\left(t_i + \tfrac{h}{2}, \mathbf{y}_i + \tfrac{h}{2}\mathbf{f}(t_i, \mathbf{y}_i)\right)$$
$$\mathbf{y}_{i+1} = \mathbf{y}_i + k_2 h$$

We can combine these together, and we will get something that looks suspiciously like the midpoint method (see equation 12.7):

$$\mathbf{y}_{i+1} = \mathbf{y}_i + \mathbf{f}\left(t_i + \tfrac{h}{2}, \mathbf{y}_i + \tfrac{h}{2}\mathbf{f}(t_i, \mathbf{y}_i)\right)h$$

The midpoint method is a second-order Runge–Kutta method. Similarly, choosing $b_1 = b_2 = \tfrac{1}{2}$ with $c_2 = a_{21} = 1$ gives us Heun's rule, which is another second-order Runge–Kutta method. Incidentally, a mathematician named Anthony Ralston came up with his own second-order Runge–Kutta method with minimized bounds on discretization error of a generic system, using $b_1 = \tfrac{1}{4}$, $b_2 = \tfrac{3}{4}$, and $c_2 = a_{21} = \tfrac{2}{3}$. As you might suspect, Ralston's RK2 will produce the same result on our example mass-spring system as the midpoint method or Heun's method [58].

The most popular Runge–Kutta method comes from Kutta's paper in 1901, and is a fourth-order approximation. The fourth-order system of equations has 11 equations and 13 unknowns, so like the second-order system, there is a family of possibilities. The most popular one, known as **Runge–Kutta 4** (or RK4) has [59]:

$$c_2 = \tfrac{1}{2} \qquad a_{21} = \tfrac{1}{2}$$
$$c_3 = \tfrac{1}{2} \qquad a_{31} = 0 \qquad a_{32} = \tfrac{1}{2}$$
$$c_4 = 1 \qquad a_{41} = 0 \qquad a_{42} = 0 \qquad a_{43} = 1$$
$$b_1 = \tfrac{1}{6} \qquad b_2 = \tfrac{1}{3} \qquad b_3 = \tfrac{1}{3} \qquad b_4 = \tfrac{1}{6}$$

This is often presented in a simplified visualization called a Butcher tableau, where the a, b, and c constants are laid out as above. The Butcher tableau for RK4 is:

$$
\begin{array}{c|cccc}
0 & & & & \\
\tfrac{1}{2} & \tfrac{1}{2} & & & \\
\tfrac{1}{2} & 0 & \tfrac{1}{2} & & \\
1 & 0 & 0 & 1 & \\
\hline
 & \tfrac{1}{6} & \tfrac{1}{3} & \tfrac{1}{3} & \tfrac{1}{6}
\end{array}
$$

Each row of the Butcher tableau corresponds to the factors used to calculate each probe vector, starting with $c_1 = 0$ for the vector k_1 (the Euler vector), indicating that we take our first probe at the start of the time window. The second row shows c_2 and a_{21}, which calculate k_2, and so on. The final row of the Butcher tableau, below the line, shows the b's used for combining the vectors.

In equation form, RK4 looks like:

$$k_1 = \mathbf{f}(t, \mathbf{y}_i)$$
$$k_2 = \mathbf{f}\left(t_i + \tfrac{h}{2}, \mathbf{y}_i + \tfrac{h}{2}k_1\right)$$
$$k_3 = \mathbf{f}\left(t_i + \tfrac{h}{2}, \mathbf{y}_i + \tfrac{h}{2}k_2\right)$$
$$k_4 = \mathbf{f}\left(t_i + h, \mathbf{y}_i + hk_3\right)$$
$$\mathbf{y}_{i+1} = \mathbf{y}_i + \tfrac{h}{6}(k_1 + 2k_2 + 2k_3 + k_4)$$

This method has a graphical interpretation shown in Figure 12.12, and there are enough zeros on the probe directions (a's) to make the geometry of it meaningful. We first cast the Euler probe and then run halfway down that line (c_2 and a_{21}) to find our next vector. We then go down that next line halfway (c_3 and a_{32}) to get our next test vector. We then get our final test vector, which we go down all the way to get our last sample. We then take a weighted average of our probe results, with the middle two probes having double the weight of the first and last ones. Intuitively, we are giving the direction of the field in the middle of the step more weight, while still accounting for the start and the end to cover curvature.

This is not the only Runge–Kutta method of order 4, although it is probably the most intuitive fourth order Runge–Kutta method. Anthony Ralston,

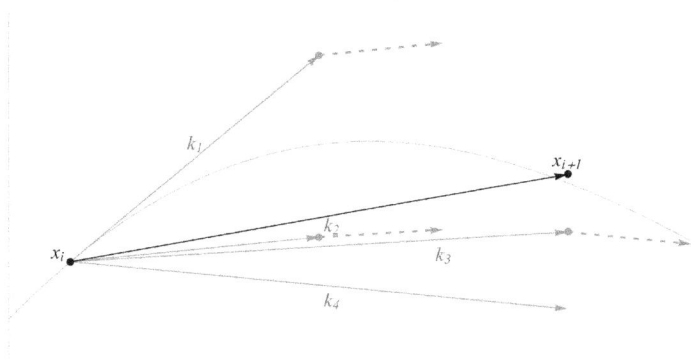

FIGURE 12.12
Geometric approach to the common Runge–Kutta 4. The four probes are sent out in order based on the result of the previous probe, and we take a weighted average of the samples.

who computed the minimum-error RK2 iteration we looked at before, also computed a form of RK4 that minimizes discretization error. Ralston's method has the Butcher tableau [58]:

$$
\begin{array}{c|cccc}
0 & & & & \\
\frac{2}{5} & \frac{2}{5} & & & \\
\frac{14-3\sqrt{5}}{16} & \frac{-2\,889+1\,428\sqrt{5}}{1\,024} & \frac{3\,785-1\,620\sqrt{5}}{1\,024} & & \\
1 & \frac{-3\,365+2\,094\sqrt{5}}{6\,040} & \frac{-975-3\,046\sqrt{5}}{2\,552} & \frac{467\,040+203\,968\sqrt{5}}{240\,845} & \\
\hline
& \frac{263+24\sqrt{5}}{1\,812} & \frac{125-1\,000\sqrt{5}}{3\,828} & \frac{3\,426\,304+1\,661\,952\sqrt{5}}{5\,924\,787} & \frac{30-4\sqrt{5}}{123}
\end{array}
$$

These are not nearly as nice as his RK2 constants were! All Runge–Kutta methods take super-linear computation time with respect to their order, but Ralston's method is even slower than the classical fourth order Runge–Kutta algorithm because there are no zero constants. Ralston's method also isn't asymptotically better—it's still fourth order—and as we saw from the mid-point/Huen comparison, Ralston's method will actually produce the same result on this simulation as the classic RK4. I would not suggest using Ralston's RK4 method at all—faster methods are available that are better—but to get some feel for what's going on here and the power of general predictor-corrector methods, a graphical representation of Ralston's RK4 is in Figure 12.13.

Ralston's RK4 method also underscores the point that Runge–Kutta methods of order n only cast n vectors, but may use a number of mathematical operations proportional to n^2: The nth vector has n parameters, which technically are tunable if you need to do so. Higher-order Runge–Kutta methods—even ones that have plenty of zeros—scale the compute cost of the integrator rapidly.

Simulations of RK4 on the mass-spring system are shown in Figure 12.14, and an accuracy comparison to leapfrog is shown in Figure 12.15. Runge–Kutta 4 is very accurate for simulating this system, beating out all prior

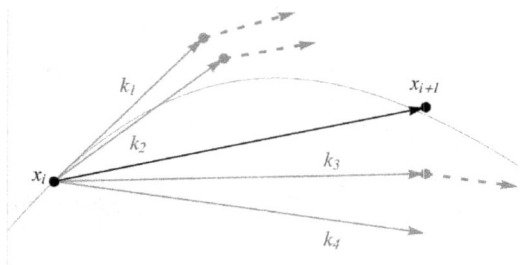

FIGURE 12.13
Geometric approach to Ralston's Runge–Kutta 4. The idea of walking down vectors to find probe points is similar to classic RK4, but the vectors and walk distances have been tuned.

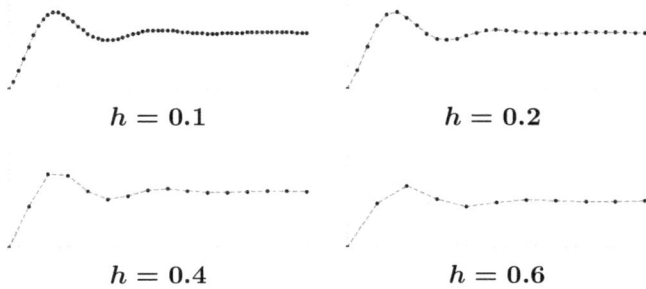

$h = 0.1$ $h = 0.2$

$h = 0.4$ $h = 0.6$

FIGURE 12.14
Runge–Kutta 4 integration of the mass-spring system from Figure 12.2 compared against the analytical solution, showing the accuracy of the simulation. RK4 tracks the correct path closely even with $h = 0.4$ and $h = 0.6$.

simulations from this chapter both for stability as h increases and accuracy with a small h. We do pay for it, though: RK4 requires us to compute and average 4 different test points to find its final simulation update. This means using more than 4 times as much compute time as a method like leapfrog integration. For some simulations, this is the appropriate balance between compute time and accuracy, but it is certainly biased toward accuracy. For higher accuracy, fifth- and sixth-order Runge–Kutta methods are available, and many of these methods have an adaptive step size to reduce overhead.

Under rounding, RK4 continues to perform well unless the time step is small, and until there are too few bits to support the subtleties of the math being done. An example of this is shown in Figure 12.16. With only five significant bits, RK4 can produce a decent simulation of the mass-spring system. However, RK4 is not energy-conserving and neither is rounding, so under heavy quantization, we can see spurious oscillations occur. Part of the reason for both the good performance and the oscillation is the final averaging step. When there are enough significant bits, this averaging step reduces quantization error by averaging it out. When there is too little significance, RK4 can

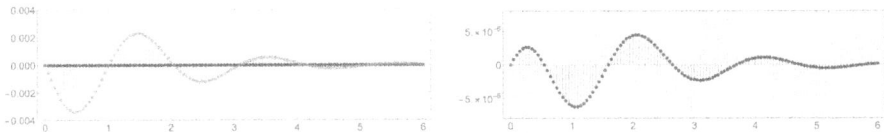

FIGURE 12.15
Error of Runge–Kutta 4 (black) compared to leapfrog integration with $h = 0.05$. The order of magnitude difference is substantially larger than the difference between leapfrog and implicit Euler.

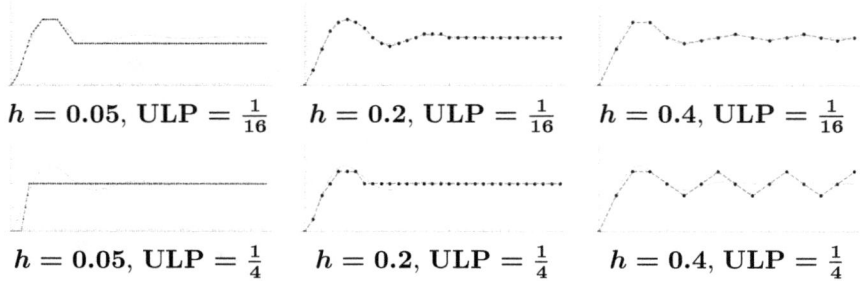

$h = 0.05$, ULP $= \frac{1}{16}$ $h = 0.2$, ULP $= \frac{1}{16}$ $h = 0.4$, ULP $= \frac{1}{16}$

$h = 0.05$, ULP $= \frac{1}{4}$ $h = 0.2$, ULP $= \frac{1}{4}$ $h = 0.4$, ULP $= \frac{1}{4}$

FIGURE 12.16
Examples of quantized simulations with Runge–Kutta 4. RK4 still has issues with high sample rates under quantization. Note that with $h = 0.4$, both simulations oscillate due to energy adding from rounding.

often oscillate because inexactness in the averaging step compounds. Quantization can also add enough error that using a high-order method gains you nothing. The example in Figure 12.17 shows that the mass-spring system has no accuracy gain over velocity Verlet until there are at least 11 bits of precision available. Even though you might prefer RK4's numerical accuracy for things like a main character's clothing, that accuracy may not be available.

If you are doing a simulation where you care about accuracy but don't need conservation of energy, it is hard to beat the Runge–Kutta methods, and RK4 in particular has a decent balance of speed and accuracy. It is also a common method to see in graphics and video game engines. Simulations of dynamic

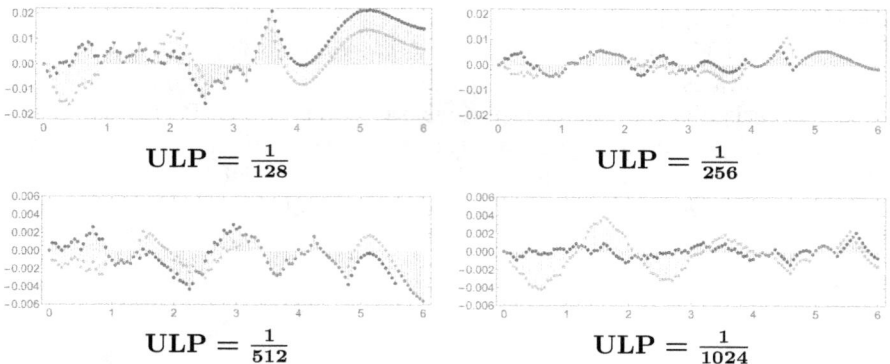

ULP $= \frac{1}{128}$ ULP $= \frac{1}{256}$

ULP $= \frac{1}{512}$ ULP $= \frac{1}{1024}$

FIGURE 12.17
Error of Runge–Kutta 4 (black) compared to leapfrog integration with $h = 0.05$ under quantization. No significant accuracy difference occurs until 11 bits of significance, and even then the difference is small compared to unquantized (see Figure 12.15). Note that the bottom two charts with over 10 bits of significance have a smaller scale than the top two.

systems with fewer speed constraints will often use high-order Runge–Kutta for short-range simulations, but switch to a symplectic method for long-range simulations where energy conservation comes more into play. However, Runge–Kutta methods are not limited to RK4 and the second-order methods. A fifth-order method with adaptive step size and a sixth-order method have gained popularity in some simulations, but the fourth-order method provides the best balance of runtime and accuracy [55].

12.7 Direct Calculation of Trajectories

So far, the methods presented in this chapter have related to the numerical simulations of a generic class of problems in differential equations called initial value problems. Initial value problems are characterized by systems of differential equations, such as Newton's laws of motion, and an initial condition. Numerical integration extends this idea in that we don't have a single initial value but a continuous input, and is unavoidable when we have things like player controls to consider. However, when we are solving initial value problems, we are not limited to approximating. If we can constrain the problem enough, we can calculate a closed-form solution ahead of time, and simply run that solution.

Projectile motion is a common situation where there is a simple closed-form solution. The projectile is launched with a specific initial velocity, and continues moving, only affected by gravity, until it hits something. We can represent projectile motion by simulating the (fixed) force of gravity on the projectile as it continues along a linear trajectory in the other dimensions, or we can simply pre-calculate the positions at each timestep using the equations for free-fall motion, with x as the distance along the horizontal dimensions and y as the vertical dimension:

$$x(t) = x_0 + v_x(t - t_0)$$
$$y(t) = y_0 + v_y(t - t_0) + \frac{1}{2}g(t - t_0)^2$$

At each timestep, we advance x and y without using the previous timestep's result. Doing the algebra, this equation corresponds perfectly with the velocity Verlet solution with continuous space, which should be expected given that this is a simulation of a second-order system with conservation of energy. However, when quantization is considered, we do not accumulate quantization error over the flight of the projectile. Instead, quantization error of x is at most one ULP, and quantization error of y is at most two ULPs throughout the simulation, and likely less. We don't mind if we have an underflow or if a term is swallowed in the sum of position, acceleration, and velocity. We will simply re-calculate the next value from scratch in the next iteration. In graphics terms, that ULP is also likely to be far less than the size of a pixel.

In this case, the closed-form solution is not any faster (or slower) to compute than the solution obtained from numerical integration, but it is more accurate. In cases of more complex motion, direct calculation can be more expensive in computational terms than the approximation. In the case of our mass-spring system, the simulation timesteps for most methods are only a few multiplications and additions, while a direct calculation for the mass-spring system requires computing two transcendental functions:

$$x(t) = 1 - \exp(-t)\cos(3t)$$

This result will likely be slow, inaccurate, or both. However, direct calculation is the only method that does not suffer from the accumulation of error over time.

12.8 Comparison of Numerical Integrators

We have discussed several different numerical integration algorithms, and their respective tradeoffs. A summary of is in Table 12.1, and a visual comparison

TABLE 12.1
Summary of common methods of numerical integration and their upsides and downsides, comparing order, RMS error at $h = 0.1$ when simulated with double-precision float, and operations per iteration to simulate a mass-spring system. Starred methods respect energy conservation. The exact number of ops used by direct calculation depends on math library.

Method	Order	Error ($h = 0.1$)	Iteration Ops
Best-known Methods			
Euler (12.1)	1	0.116	4
Semi-implicit Euler* (12.2)	1	0.037	4
Velocity Verlet* (12.4, 12.5)	2	0.005	7
Hacked velocity Verlet (12.6)	1	0.044	6
Midpoint (12.7)	2	0.008	8
Runge–Kutta 4	4	0.00004	22
Less Common Methods			
Leapfrog* (12.3)	2	0.005	4
Heun (12.8)	2	0.008	8
Ralston RK2	2	0.008	12
Ralston RK4	4	0.00004	35
Direct Calculation*	∞	Near 0	30–60

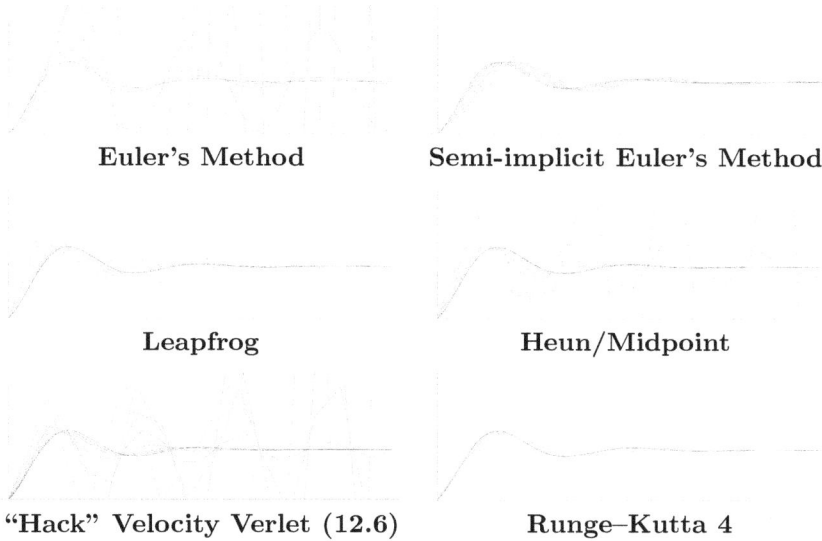

Euler's Method   Semi-implicit Euler's Method

Leapfrog   Heun/Midpoint

"Hack" Velocity Verlet (12.6)   Runge–Kutta 4

FIGURE 12.18
Visual comparison of mass-spring system simulations with h going from 0.05 to 0.6. First-order methods are on the first row, second-order methods on the second row, and two other methods are on the last row: Velocity Verlet with the explicit velocity hack (Equation 12.6) and RK4.

of these algorithms on a damped mass-spring system is shown in Figure 12.18. A plot showing their error as tick size changes is given in Figure 12.19.

We started with Euler's method, the naive approach to numerical integration, and noted how it needed a high simulation frame rate to even converge, let alone producing accurate results, and then we modified it to compute one term implicitly, generating an algorithm that is more stable and performs well on periodic simulations. We also saw the tradeoff between discretization and quantization error that occurs when setting the frame rate, and how high frame rates can hurt simulation accuracy when quantization comes into play.

From Euler, we looked at ways to perform updates at the midpoints between our calculations to create second-order approximations. When we specialize around having an equation for position and an equation for velocity, we can leapfrog between them, creating the second-order methods of leapfrog integration and velocity Verlet. When we look more generally, we can attempt to use Euler-like steps to predict possible next values of our simulation, and compute our actual next value based on the midpoint of that prediction or the average of multiple predictions. Finally, we extended the process of prediction and correction to a set of higher-order methods that have very high numerical

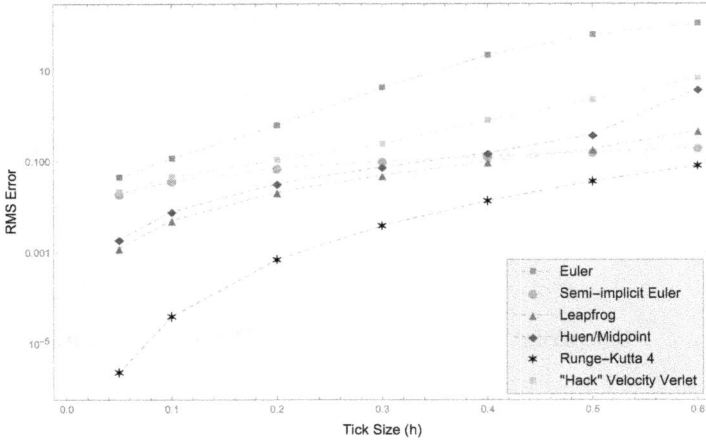

FIGURE 12.19
Log-scale plot of root-mean-squared error of each simulation run on our mass-spring system across all the methods discussed in this chapter. At low tick size, accuracy improves by orders of magnitude as order of the numerical method increases, and at higher tick size, stability becomes more important.

accuracy. We finished the discussion with a small discussion of direct calculation, an alternative to simulation when we have a simple closed form.

In some of these cases, specifically the time-staggered approaches of leapfrog, velocity Verlet, and semi-implicit Euler, we needed to do some algebra to account for velocity-dependent acceleration. In some contexts, this limitation may preclude these methods if this algebra needs to be done programmatically. Similarly, these methods cannot be adapted to simulate third-order or higher-order differential equations. The Runge–Kutta family of methods are more generally applicable.

Referring back to Table 12.1 and Figure 12.18, semi-implicit Euler and leapfrog integration outperformed other algorithms in terms of the speed/accuracy tradeoff on our example periodic system. However, Runge–Kutta 4 still had the best simulation accuracy overall. We can also see this comparison in another way in the root-mean-squared error of the simulations, shown in Figure 12.19. With a fast simulation frame rate, RK4's accuracy was pixel-perfect, even when zoomed in, while the second-order methods had RMS error of around 10^{-4} and the first-order methods had error near 10^{-2}.

We will end with a final visual comparison, in Figure 12.20, of an eccentric orbit of a planet around a sun. All numerical integration methods have a failure mode here compared to a direct calculation. Error in the non-symplectic integrators manifests as energy gain or loss in the system, resulting in the planet either shooting off into space or crashing into the sun. Error in the symplectic integrators manifests in an orbital procession, where the orbit's

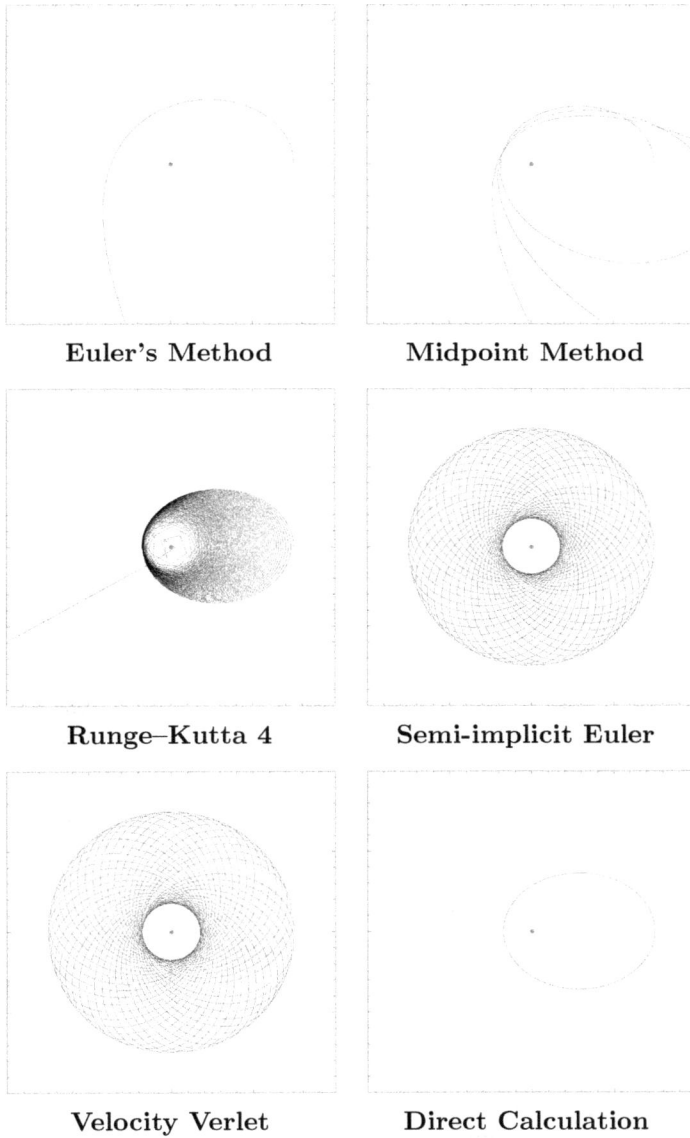

FIGURE 12.20

Visual comparison of an eccentric orbit as simulated with the methods from this chapter. Aside from direct calculation, every method has a failure mode: Euler and Midpoint add energy to the planet, causing it to break orbit and fly off. RK4 subtracts energy (very slowly), causing the planet to crash into the sun and finally become unstable. The symplectic integrators cause a spurious procession of the orbit, while maintaining orbital altitude.

altitude stays the same, but it slowly rotates around the sun. This is not a quantization artifact either: The simulations in Figure 12.20 produce nearly identical results down to relatively low precision. In many cases, with numerical integration, you will simply have to pick your desired failure modes.

General Recommendations for Numerical Integration

- For accuracy, prefer velocity Verlet or leapfrog integration for long simulations and RK4 for short simulations. Consider higher-order methods (e.g., fourth order symplectic or sixth order Runge–Kutta) if these are insufficient.

- For speed, prefer leapfrog integration for the best tradeoffs of speed and accuracy or alternatively consider semi-implicit Euler.

- When specializing for Newtonian physics, there are more accurate algorithms the same speed as Euler, and there are equally good algorithms that are faster than midpoint.

- Position-based dynamics greatly simplify the math involved in the semi-implicit methods, including leapfrog and velocity Verlet.

- Higher-order methods with wide step size get the same accuracy as lower-order methods with smaller step size.

- For faraway or heavily quantized simulations, a larger simulation tick size may increase accuracy, and a higher-order algorithm probably won't.

Check Your Understanding

Problem 12.1. Simulate a predator-prey system. The predators prefer to stay away from each other and are attracted toward prey. The prey prefer to run from the predators. Have the predators eat any prey that is within a radius of them, and have the predators die if they haven't eaten in a certain amount of time. Model each of these preferences as a force on the predators and the prey, but also cap the speed of the predators and the prey. Once you have your system, simulate with a numerical method of your choice. Find conditions where the predators eat all the prey. Find conditions where the prey outlast the predators.

    Consider also adding: Herd behavior for prey, impassable obstacles, or food sources for prey.

Problem 12.2. Make a simulator for a cannonball shot out of a cannon with Euler integration and single-precision floating point. Move the cannon toward $x = \infty$ until you get a result you didn't expect. Vary simulation frame rate until you get a result that you didn't expect. Use a second-order method and see where your error conditions continue.

Problem 12.3. Make a simulation of a rope bridge accounting for the weight of the planks of the bridge and the tension of the rope between planks (do a 2D side view to make this simpler). Drop a ball on the bridge, and simulate the collision and forces from the collision. Try making the planks of the bridge lighter and see what happens to the simulation.

Problem 12.4. Make a simulator for planetary orbits around a star using at least a symplectic integrator of your choice and a predictor-corrector method, recreating the results of Figure 12.20. Try a circular orbit and an eccentric orbit. Find the values of h for each algorithm where the orbit looks visually stable.

Add a direct calculation of an orbit to your simulator (*Hint*: Use polar coordinates and follow Kepler's equations of motion). Compare the results.

Problem 12.5. Come up with a scheme for a directly computed double jump in a 2D platforming game, modeling each jump as an instantaneous upward force, with the double jump canceling downward momentum. Come up with a scheme for numerical calculation of the same jump using a first- or second-order method. Which scheme performs faster? Which looks better?

Some games add variable gravity, where gravity increases as a player descends. Incorporate that into your scheme. Compare numerical methods to a direct calculation.

Problem 12.6. Implement a simulation of a triple pendulum using RK4 and velocity Verlet. Give the pendulum small perturbations. Do the results match your expectations? Try giving larger and larger permutations until the pendulum does something unintuitive.

Bibliography

[1] "IEEE Standard for Floating-Point Arithmetic," *IEEE Std 754-2019 (Revision of IEEE 754-2008)*, pp. 1–84, 2019. DOI: `10.1109/IEEESTD.2019.8766229`.

[2] R. Rojas, "Konrad zuse's legacy: The architecture of the z1 and z3," *IEEE Annals of the History of Computing*, vol. 19, no. 2, pp. 5–16, 1997. DOI: `10.1109/85.586067`.

[3] Intel, *Intel® 64 and ia-32 architectures software developer's manual*, 2024. [Online]. Available: `https://www.intel.com/content/www/us/en/developer/articles/technical/intel-sdm.html`.

[4] C. Severance, *An interview with the old man of floating-point*, 1998. [Online]. Available: `https://people.eecs.berkeley.edu/~wkahan/ieee754status/754story.html`.

[5] J. Sohn, D. K. Dean, E. Quintana, and W. S. Wong, "Enhanced floating-point multiply-add with full denormal support," in *2023 IEEE 30th Symposium on Computer Arithmetic (ARITH)*, 2023, pp. 143–150. DOI: `10.1109/ARITH58626.2023.00015`.

[6] P. Duperas, *Nan boxing or how to make the world dynamic*, 2020. [Online]. Available: `https://piotrduperas.com/posts/nan-boxing`.

[7] Arm, *Arm architecture reference manual for a-profile architecture*, 2024. [Online]. Available: `https://developer.arm.com/documentation/ddi0487/ka/?lang=en`.

[8] ECMA, *ECMA-262: ECMAScript Language Specification*, 15th ed. Geneva, Switzerland: ECMA (European Association for Standardizing Information and Communication Systems), 2024. [Online]. Available: `https://ecma-international.org/wp-content/uploads/ECMA-262_15th_edition_june_2024.pdf`.

[9] B. D. Rouhani, R. Zhao, A. More, *et al.*, *Microscaling data formats for deep learning*, 2023. arXiv: `2310.10537 [cs.LG]`. [Online]. Available: `https://arxiv.org/abs/2310.10537`.

[10] IBM, *IBM z16 (3931) technical guide*, IBM Redbooks, 2024. [Online]. Available: `https://www.redbooks.ibm.com/redbooks/pdfs/sg248951.pdf`.

[11] D. Goldberg, "What every computer scientist should know about floating-point arithmetic," *ACM Computing Surveys*, vol. 23, no. 1, 5–48, 1991, ISSN: 0360-0300. DOI: 10.1145/103162.103163. [Online]. Available: https://doi.org/10.1145/103162.103163.

[12] NVIDIA, *Floating point and ieee 754 compliance for nvidia gpus*, 2024. [Online]. Available: https://docs.nvidia.com/cuda/floating-point/index.html.

[13] J. D. Bruguera, "Low-latency and high-bandwidth pipelined radix-64 division and square root unit," in *2022 IEEE 29th Symposium on Computer Arithmetic (ARITH)*, 2022, pp. 10–17. DOI: 10.1109/ARITH54963.2022.00012.

[14] N. Badizadegan, "Newton-raphson integer division for area-constrained microcontrollers," in *2023 IEEE 30th Symposium on Computer Arithmetic (ARITH)*, 2023, pp. 9–15. DOI: 10.1109/ARITH58626.2023.00024.

[15] M. Flynn, "On division by functional iteration," *IEEE Transactions on Computers*, vol. C-19, no. 8, pp. 702–706, 1970. DOI: 10.1109/T-C.1970.223019.

[16] "Information technology – Programming languages – C," International Organization for Standardization, Geneva, CH, Standard ISO/IEC 9899:2024, 2023.

[17] Nvidia, *NVIDIA nsight visual studio edition: Achieved FLOPs*, 2015. [Online]. Available: https://docs.nvidia.com/gameworks/content/developertools/desktop/analysis/report/cudaexperiments/kernellevel/achievedflops.htm.

[18] R. Sommefeldt, *Boats crash/break and can kill their passengers when falling certain distances*, 2006. [Online]. Available: https://www.beyond3d.com/content/articles/15/.

[19] J. Lyons, *ARIANE 5 flight 501 failure*, 1996. [Online]. Available: https://web.archive.org/web/20000815230639/http://www.esrin.esa.it/htdocs/tidc/Press/Press96/ariane5rep.html.

[20] B. Gladman, V. Innocente, J. Mather, and P. Zimmermann, *Accuracy of mathematical functions in single, double, double extended, and quadruple precision*, 2024. [Online]. Available: https://members.loria.fr/PZimmermann/papers/accuracy.pdf.

[21] J. L. Spouge, *Computation of the Gamma, Digamma, and Trigamma Functions*, Apr. 2018. DOI: 10.1137/0731050. [Online]. Available: https://doi.org/10.1137/0731050.

[22] WebKit, *Class jsvalue (lines 413–471)*, 2024. [Online]. Available: https://github.com/WebKit/WebKit/blob/safari-7619.1.14-branch/Source/JavaScriptCore/runtime/JSCJSValue.h.

[23] P. Panchekha, A. Sanchez-Stern, J. R. Wilcox, and Z. Tatlock, "Automatically improving accuracy for floating point expressions," in *Proceedings of the 36th ACM SIGPLAN Conference on Programming Language Design and Implementation*, ser. PLDI '15, Portland, OR, USA: Association for Computing Machinery, 2015, 1–11, ISBN: 9781450334686. DOI: 10.1145/2737924.2737959. [Online]. Available: https://doi.org/10.1145/2737924.2737959.

[24] J.-M. Muller, N. Brisebarre, F. de Dinechin, *et al.*, *Handbook of Floating-Point Arithmetic*. Cham, Switzerland: Birkhäuser, 2018, ISBN: 978-3-319-76525-9. DOI: 10.1007/978-3-319-76525-9.

[25] "PATRIOT MISSILE DEFENSE: software problem led to system failure at Dhahran, Saudi Arabia," United States General Accounting Office, Washington DC, US, Tech. Rep. GAO/IMTEC-92-26, 1992.

[26] A. G. Stephenson, L. S. LaPiana, D. R. Mulville, *et al.*, "Mars climate orbiter mishap investigation board phase i report," NASA, Tech. Rep., 1999.

[27] M. Pusz, *Mp-units – the quantities and units library for C++*, 2023. [Online]. Available: https://github.com/mpusz/mp-units.

[28] Y. Hida, S. Li, and D. Bailey, "Library for double-double and quad-double arithmetic," Jan. 2008.

[29] M. Joldes, J.-M. Muller, and V. Popescu, "Tight and rigorous error bounds for basic building blocks of double-word arithmetic," *ACM Transactions on Mathematical Software*, vol. 44, no. 2, 2017, ISSN: 0098-3500. DOI: 10.1145/3121432. [Online]. Available: https://doi.org/10.1145/3121432.

[30] N. J. Higham, "The accuracy of floating point summation," *SIAM Journal on Scientific Computing*, vol. 14, no. 4, pp. 783–799, 1993. DOI: 10.1137/0914050. eprint: https://doi.org/10.1137/0914050. [Online]. Available: https://doi.org/10.1137/0914050.

[31] W. Ahrens, J. Demmel, and H. D. Nguyen, "Algorithms for efficient reproducible floating point summation," *ACM Transactions on Mathematical Software*, vol. 46, no. 3, 2020, ISSN: 0098-3500. DOI: 10.1145/3389360. [Online]. Available: https://doi.org/10.1145/3389360.

[32] F. S. Foundation, *3.11 options that control optimization*, 2024. [Online]. Available: https://gcc.gnu.org/onlinedocs/gcc/Optimize-Options.html.

[33] J. Mason and D. Handscomb, *Chebyshev Polynomials*. New York: Chapman and Hall/CRC, 2003, ISBN: 978-1-4200-3611-4. DOI: 10.1201/9781420036114.

[34] W. Fraser, "A survey of methods of computing minimax and near-minimax polynomial approximations for functions of a single independent variable," *Journal of the ACM*, vol. 12, no. 3, 295–314, 1965, ISSN: 0004-5411.

[35] J. P. Lim, "Novel polynomial approximation methods for generating correctly rounded elementary functions," Ph.D. dissertation, Rutgers University, 2021. DOI: 10.7282/t3-hn8j-qx87.

[36] N. Brisebarre and S. Chevillard, "Efficient polynomial l-approximations," in *18th IEEE Symposium on Computer Arithmetic (ARITH '07)*, 2007, pp. 169–176. DOI: 10.1109/ARITH.2007.17.

[37] A. H. Karp and P. Markstein, "High-precision division and square root," *ACM Transactions on Mathematical Software*, vol. 23, no. 4, 561–589, 1997, ISSN: 0098-3500. DOI: 10.1145/279232.279237. [Online]. Available: https://doi.org/10.1145/279232.279237.

[38] G. Cozma and C. Lam, *AMD's Zen 4, part 2: Memory subsystem and conclusion*, 2022. [Online]. Available: https://chipsandcheese.com/2022/11/08/amds-zen-4-part-2-memory-subsystem-and-conclusion/.

[39] C. Lam, *Previewing Meteor Lake at CES*, 2024. [Online]. Available: https://chipsandcheese.com/2024/01/11/previewing-meteor-lake-at-ces/.

[40] N. Badizadegan, *You (probably) shouldn't use a lookup table*, 2022. [Online]. Available: https://specbranch.com/posts/lookup-tables/.

[41] O. Desrentes and F. de Dinechin, "Using integer linear programming for correctly rounded multipartite architectures," in *2022 International Conference on Field-Programmable Technology (ICFPT)*, 2022, pp. 1–8. DOI: 10.1109/ICFPT56656.2022.9974486.

[42] Docper, *Origin of quake3's fast invsqrt() – part two*, 2017. [Online]. Available: https://bugs.mojang.com/browse/MC-119369.

[43] C. Lomont, *Fast inverse square root*, 2003. [Online]. Available: https://www.lomont.org/papers/2003/InvSqrt.pdf.

[44] R. Gupta, "Bhāskara i's approximation to sine," *Indian Journal of History of Science*, vol. 2, no. 2, pp. 121–136, 1967. [Online]. Available: https://web.archive.org/web/20120316083451/http://www.new.dli.ernet.in/rawdataupload/upload/insa/INSA_1/20005af0_121.pdf.

[45] W. J. Gordon and R. F. Riesenfeld, "Bernstein-bézier methods for the computer-aided design of free-form curves and surfaces," *Journal of the ACM*, vol. 21, no. 2, 293–310, Apr. 1974, ISSN: 0004-5411. DOI: 10.1145/321812.321824. [Online]. Available: https://doi.org/10.1145/321812.321824.

[46] L. Piegl and W. Tiller, *The NURBS Book*, 2nd ed. Springer Berlin, Heidlberg, 1996, ISBN: 9783540615453. DOI: 10.1007/978-3-642-59223-2.

[47] T. Akenine-Möller, E. Haines, N. Hoffman, A. Pesce, M. Iwanicki, and S. Hillaire, *Real-Time Rendering 4th Edition*. Boca Raton, FL, USA: A K Peters/CRC Press, 2018, p. 1200, ISBN: 9781138627000.

[48] E. Persson and A. Studios, "Creating vast game worlds: Experiences from avalanche studios," in *ACM SIGGRAPH 2012 Talks*, ser. SIG-GRAPH '12, Los Angeles, California: Association for Computing Machinery, 2012, ISBN: 9781450316835. DOI: 10.1145/2343045.2343089. [Online]. Available: https://doi.org/10.1145/2343045.2343089.

[49] J. Gregory, *Game Engine Architecture Third Edition*. Boca Raton, FL, USA: CRC Press, 2019, ISBN: 9781138035454.

[50] E. Gilbert, D. Johnson, and S. Keerthi, "A fast procedure for computing the distance between complex objects in three-dimensional space," *IEEE Journal on Robotics and Automation*, vol. 4, no. 2, pp. 193–203, 1988. DOI: 10.1109/56.2083.

[51] M. Montanari, N. Petrinic, and E. Barbieri, "Improving the gjk algorithm for faster and more reliable distance queries between convex objects," *ACM Transactions on Graphics*, vol. 36, no. 3, Jun. 2017, ISSN: 0730-0301. DOI: 10.1145/3083724. [Online]. Available: https://doi.org/10.1145/3083724.

[52] J. E. Bresenham, "Algorithm for computer control of a digital plotter," *IBM Systems Journal*, vol. 4, no. 1, pp. 25–30, 1965. DOI: 10.1147/sj.41.0025.

[53] L. F. Shampine and C. W. Gear, "A user's view of solving stiff ordinary differential equations," *SIAM Review*, vol. 21, no. 1, pp. 1–17, 1979, ISSN: 00361445. [Online]. Available: http://www.jstor.org/stable/2029929 (visited on 07/31/2024).

[54] *Numerical Differential Equation Methods*. John Wiley & Sons, Ltd, 2016, ISBN: 9781119121534. DOI: https://doi.org/10.1002/9781119121534.ch2. eprint: https://onlinelibrary.wiley.com/doi/pdf/10.1002/9781119121534.ch2. [Online]. Available: https://onlinelibrary.wiley.com/doi/abs/10.1002/9781119121534.ch2.

[55] W. H. Press, S. A. Teukolsky, W. T. Vetterling, and B. P. Flannery, *Numerical Recipes 3rd Edition: The Art of Scientific Computing*, 3rd ed. Cambridge University Press, 2007, ISBN: 0521880688. [Online]. Available: http://www.amazon.com/Numerical-Recipes-3rd-Scientific-Computing/dp/0521880688/ref=sr_1_1?ie=UTF8&s=books&qid=1280322496&sr=8-1.

[56] M. Müller, B. Heidelberger, M. Hennix, and J. Ratcliff, "Position based dynamics," *Journal of Visual Communication and Image Representation*, vol. 18, no. 2, 109–118, Apr. 2007, ISSN: 1047-3203. DOI: 10.1016/ j.jvcir.2007.01.005. [Online]. Available: https://doi.org/10. 1016/j.jvcir.2007.01.005.

[57] Wikipedia, *Verlet integration: Algorithmic representation*, 2024. [Online]. Available: https://en.wikipedia.org/wiki/Verlet_ integration#Algorithmic_representation.

[58] A. Ralston, "Runge-kutta methods with minimum error bounds," *Mathematics of Computation*, vol. 16, pp. 431–437, 1962. [Online]. Available: https://api.semanticscholar.org/CorpusID:122669148.

[59] W. Kutta, "Beitrag zur näherungsweisen Integration totaler Differentialgleichungen," *Zeitschrift für Mathematik und Physik*, vol. 46, pp. 435–53, 1901.

Index

For Product Safety Concerns and Information please contact our EU
representative GPSR@taylorandfrancis.com
Taylor & Francis Verlag GmbH, Kaufingerstraße 24, 80331 München, Germany

www.ingramcontent.com/pod-product-compliance
Lightning Source LLC
Chambersburg PA
CBHW060551220326
41598CB00024B/3074